INSIDE ArcCAD®

Michael A. DeLacey

ONWORD® PRESS

INSIDE ArcCAD®

By Michael A. DeLacey

Published by:

OnWord Press
2530 Camino Entrada
Santa Fe, NM 87505-4835 USA

SAN 694-0269
First Edition, 1995

10 9 8 7 6 5 4 3 2 1

Printed in the United States of America

Library of Congress Cataloging-in-Publication Data

DeLacey, Michael A.
INSIDE ArcCAD / Michael A. Delacey.
p. cm.
Includes index.
ISBN 1-56690-084-0
1. Geographic information systems 2. ArcCAD.
I. Title
G70.2.D45 1995
910'.285—dc20 95-33179
CIP

Trademarks

Warning and Disclaimer

About the Author

Michael A. DeLacey is the president and co-founder of Microdesk of New England, Inc., an Autodesk VAR and consulting firm located in Waltham, MA. Prior to founding Microdesk, Mr. DeLacey worked for three years as an applications engineer for an Autodesk reseller and five years as a CAD manager for a civil/survey firm.

Acknowledgments

I would like to thank David Talbott of OnWord Press for providing me with the opportunity to write this book and Margaret Burns, my editor, also of OnWord Press, for seeing that I actually completed it and for having patience with me during the development. The technical team at ESRI also deserves to be commended for the speed with which they provided beta copies of ArcCAD and the way they handled all of my technical questions.

The data sets used throughout this book were provided by David Strebb and the City of Fitchburg, MA. The case studies were provided by the following: Michael Vignale of Beta Engineering, Lincoln, RI, who provided the RI Traffic Management Systems case study; and David Caruth, Senior Transportation Engineer, and Edward M. Conover, Chief Traffic Engineer, of The Port Authority of New York and New Jersey who provided the Newark Airport Traffic Management System case study. I would also like to thank my office and in particular my partner Robin Adams for taking care of the business while my attentions were concentrated on this project.

Last but not least, I would like to thank my wife Cathleen for putting up with me or at least my absence over the last several months. I have been able to live my life confident in the fact that no matter what I try I have two people behind me with complete faith and support. It is for this reason that I dedicate this book to my parents Lionel and Genavieve DeLacey.

Book Production

This book was produced using Ventura 4.1.1 Desktop Publishing software running on 80486 PCs. Text was produced with a variety of word processing programs. The cover was produced by Lynne Egensteiner, using Quark Xpress 3.11 and Adobe Photoshop 2.5.1 on a Mac Quadra.

OnWord Press

OnWord Press is dedicated to the fine art of professional documentation.

In addition to the author who developed the material for this book, other members of the OnWord Press team make the book end up in your hands. In addition to those listed below, thanks to Joe Adams, James Bridge, Roxsan Meyer, Jean Nichols, Zeke Olguin, Robin Ortiz, and Lisa Meyrick.

Dan Raker, Publisher and President
Gary Lange, Vice President
David Talbott, Acquisitions Editor
Margaret Burns, Senior Editor
Carol Leyba, Production Manager
Janet Leigh Dick, Marketing Director
Kate Bemis, Indexer
Lynne Egensteiner, Cover Designer

Contents

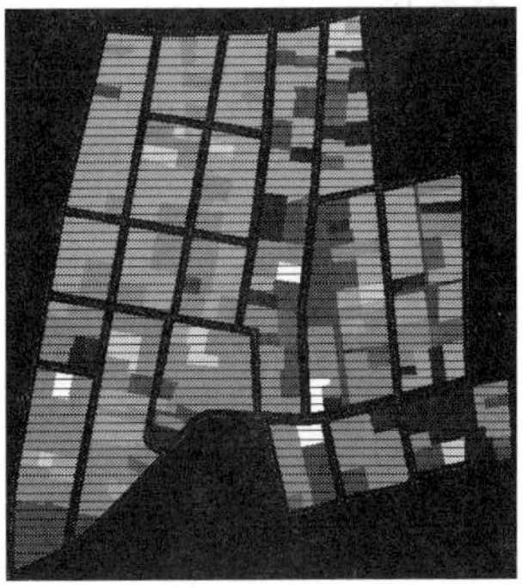

Introduction

ArcCAD, developed by Environmental Systems Research Institute (ESRI), is a PC-based geographic information systems (GIS) toolkit for use within an AutoCAD computer-aided design (CAD) system. Prior to ArcCAD, there were no inexpensive, relatively simple GIS options. In 1989, Autodesk and ESRI joined forces to develop a package to bring GIS to the CAD market. States and towns were searching for a better method to track and maintain their resources. End users within these government entities were demanding an easy-to-use system that would allow them to gather, store, analyze, and display their geographic data.

With the advent of ArcCAD, end users were able to use the familiar AutoCAD interface to manage their GIS tasks. With this new low-cost, desktop system, many other end users have begun to look at GIS as an invaluable tool for many diverse market segments. Today, ArcCAD is used for a wide range of applications. Users include environmental, civil, and transportation engineering firms, marketing research firms, business, and real estate companies. As the mapping data being produced by states and towns becomes readily accessible, the number of users is expected to grow considerably.

What This Book Is For

INSIDE ArcCAD is a tutorial and reference for ArcCAD. The book provides the new user of ArcCAD with the ability to get up and running quickly and the experienced user with a supplemental reference.

This book focuses on ArcCAD's commands to create, analyze, and display GIS datasets. After completing the exercises, you will have the knowledge to create new datasets, access existing datasets, create spatial overlays, thematic maps, and generate statistical analysis of GIS data.

What This Book Is Not For

INSIDE ArcCAD is not meant to teach GIS or AutoCAD. It is meant to be a supplement to your AutoCAD command references and your ArcCAD manuals. Throughout the book there are references to basic GIS principles and AutoCAD techniques. If you are a new user of AutoCAD, we recommend you become familiar with AutoCAD through the reference manuals, tutorials, and the numerous third-party books currently available.

How the Book Is Organized

Part I. Building Geographic Data

- Assembling spatial data
- Correcting spatial data
- Creating topology
- Creating and using databases

Part II. Analyzing Data

- Preparing data for analysis
- Spatial analysis
- Tabular queries

Part III. Creating Maps

- Displaying data
- Intelligent maps

Part IV. Case Studies

What You Need To Use This Book

To work with *INSIDE ArcCAD* you will need a system capable of running AutoCAD Release 13 with ArcCAD Release 11.4. AutoCAD and ArcCAD should be installed and fully configured, including setting the AutoCAD rendering environment (AVE). You will need at least 10MB of available disk space on your hard drive. Be sure that AutoCAD and ArcCAD are running properly before you begin to work with this book.

If you have difficulty installing or configuring either AutoCAD or ArcCAD, check your documentation or contact your authorized AutoCAD or ArcCAD dealer.

DOS Versus Windows

INSIDE ArcCAD was written for use with the Windows version of ArcCAD. Because the DOS and Windows versions of ArcCAD are essentially the same, *INSIDE ArcCAD* is equally valuable for users working within DOS.

Installing the Data Disk

To install the tutorial data, follow these steps:

In DOS, change to the root directory of your C drive and enter the following:

```
C:Xcopy a: c: /s
```

In Windows, use the File Manager to drag the INARC folder onto the C drive.

By doing either of the above commands you will have copied all of the tutorial data into the INARC directory on your C drive. The tutorial data contains spatial data, tabular databases, and key files for use throughout the book. It is important that the information be placed in the C:\INARC directory. If the data is not placed in this directory, ArcCAD will not be able to find the information when needed.

Once the data has been copied to your hard drive, be sure to make the INARC directory the working directory before entering ArcCAD. To be sure that INARC is the working directory in DOS, change to the INARC directory before executing the ArcCAD batch file. In Windows, set the working directory edit box in the properties dialog box to C:\INARC.

Part I.

Building Geographic Data

Chapter

Overview of ArcCAD

What Is GIS?

The definition of GIS as found in most textbooks is as follows:

> An organized collection of computer hardware, software, geographic data, and personnel designed to efficiently capture, store, update, manipulate, analyze, and display all forms of geographically referenced information.

This book attempts to simplify this definition by providing examples and real-world applications of how ArcCAD can be used to create, analyze, and display GIS data.

We assume you have assembled the necessary hardware and have installed and configured both the AutoCAD and ArcCAD software products. The next step is to begin compiling the data. Data can be in the form of CAD drawings, paper drawings, scanned images, text documents, spreadsheet data files, database files, and any form of data that is used to represent geographic entities.

Once the data has been compiled, the last step in forming a GIS is to train or hire personnel to manipulate, analyze, and most importantly, display the geographic information in a usable form. For a GIS to be successful and functional, end users must have the ability to do complex analysis and be able to output the results.

Applying GIS

This section presents a sample application of a GIS and ArcCAD in a real-world scenario. Take some time to read the following section to get a feel for GIS and how the concepts presented might be applied to your own situation.

John Smith was recently promoted to Assistant Town Manager after 14 years of service. With this promotion comes a mixture of excitement and trepidation. John is struck by the magnitude of his first assignment. Within a week, a site must be chosen for the new elementary school. John begins to gather information by looking at maps the town has prepared.

The town's Planning Department has an AutoCAD system where they maintain the parcel maps for tax mapping. With a few phone calls, John finds that the County Environmental Protection Agency uses a PC ARC/INFO system to maintain all hydrological features, including lakes, ponds, rivers, streams, wetlands, and flood plains. He also is directed to a national data provider who can provide the most recent census data in an ARC/INFO format, which would include owner names, number of children per household, and average income. John realizes that to accomplish the task at hand the town needs a way to combine, compile, and analyze all of this different data to identify the best site for the new school.

After careful evaluation, John decides on a software product called ArcCAD. Because ArcCAD is an AutoCAD application, it can easily make use of the AutoCAD drawing files. Furthermore, because ArcCAD is an ESRI product, it can directly read the PC ARC/INFO coverages available from the Environmental Protection Agency and the ARC/INFO census information delivered by the national data provider.

With all of the data compiled, John first generates a quick thematic map identifying parcels owned by families with three or more children in bright red and all other parcels in green. This allows him to intelligently narrow his study area and limit the bulk of his analysis. John then uses ArcCAD's query capabilities to spatially select all town-owned lands within 5 miles of the study area. He then performs an additional query to further limit the parcels to those larger than the required 5 acre minimum.

With this group of potential parcels, John begins to further analyze the appropriateness of the land. His experience has shown that any land within 300 feet of a flood plain or wetland area will not be acceptable. John uses ArcCAD's buffering capabilities to generate a corridor around all wetlands and flood plains within his study area. Using ArcCAD's overlay functions, John then intersects this corridor buffer with the EPA's hydrological information and the likely parcels to form a new dataset.

The resulting intersection identifies town-owned parcels that are not within the wetlands and flood plains. One more query, selecting parcels that are larger than 5 acres that are outside the corridor buffer, leaves John with a limited choice of possible locations for the new school. In addition to the parcels listing, John also has all of the thematic maps, datasets, and reports necessary to clearly and concisely present his findings to the Planning Board.

This story is a simplified version of the process that may be used to locate a new school within a town, but it does depict the capabilities of the ArcCAD GIS toolkit. Throughout *INSIDE ArcCAD* you will use examples like this to learn and use ArcCAD.

What Is ArcCAD?

ArcCAD is a set of tools developed by Environmental Systems Research Institute, Inc. (ESRI) for creating and working with spatial and tabular data within Autodesk's AutoCAD drafting environment. AutoCAD, as the industry leader in PC-based computer-aided design (CAD) software, has over one million licenses in use and is taught to over one million students around the world each year. By developing ArcCAD to work within the AutoCAD environment, ESRI has provided a comprehensive set of GIS tools available to millions of users around the world.

ArcCAD is available to run on the DOS, Windows, and Windows NT platforms. Because it is an application running within AutoCAD, ArcCAD supports all of the peripheral products and interfaces supported by AutoCAD. ArcCAD conforms to AutoCAD's user interface and can readily take advantage of all data available in either the AutoCAD Drawing (DWG) or Data Interchange File (DXF) formats.

ArcCAD is one of a suite of GIS products produced by ESRI. ARC/INFO, ESRI's flagship product, is a high-end application for use on UNIX and VMS workstations. ARC/INFO is widely used around the world in many state and county-wide GIS programs, as well as by GIS consultants and data providers. ARC/INFO provides some complex functions not yet available within ArcCAD, such as network tracing, dynamic segmentation, node attribute tables, and event tables.

ESRI's PC ARC/INFO is a subset of the ARC/INFO toolset ported to the PC platform. PC ARC/INFO has a comprehensive set of GIS functions, but uses its own editing environment instead of AutoCAD's.

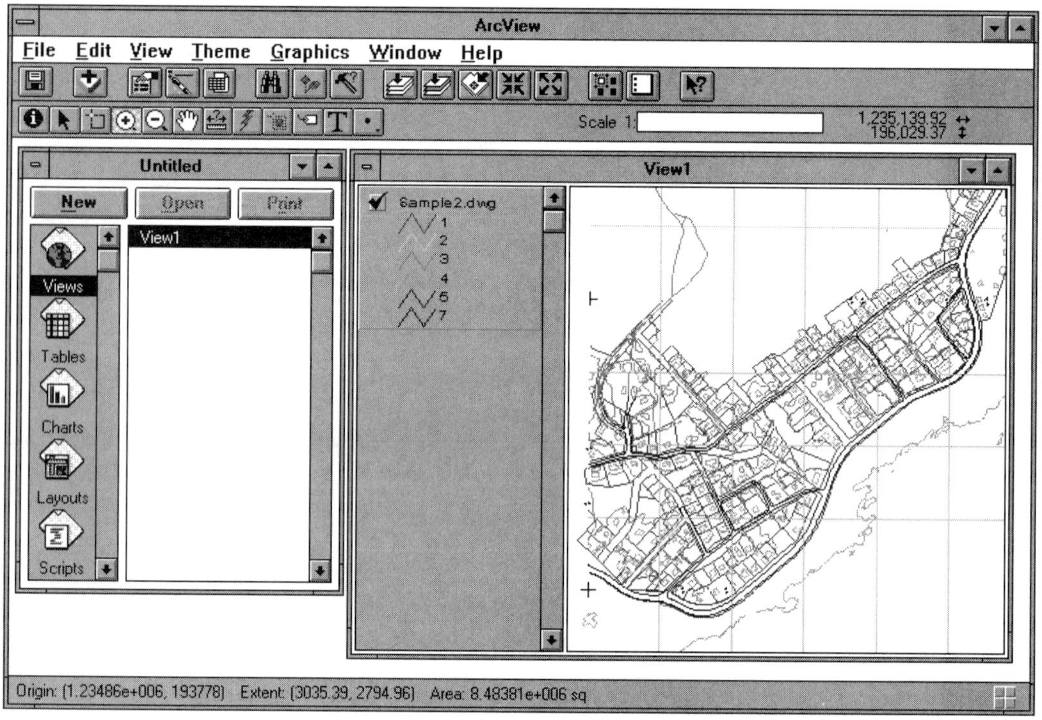

The ArcView interface.

ArcView is ESRI's latest GIS product offering. ArcView is a multiplatform software package for querying and displaying GIS data created with ARC/INFO, PC ARC/INFO, or ArcCAD. ArcView is positioned to bring the display and use of GIS data to every desktop. This package can provide the manager or casual GIS user with the ability to do complex analysis on geographic data with minimal software or GIS experience.

Throughout its product line, ESRI has maintained compatibility so that data created can be shared across products and platforms. This enables ArcCAD users to use data created by any of ESRI's other products.

For more information on products available from ESRI, contact ESRI at 909/793-2853.

How ArcCAD Works

ArcCAD facilitates the gathering, storing, analysis, and display of geographic data by converting graphics, AutoCAD drawings, Data Interchange Files, PC ARC/INFO data, and ARC/INFO data into a special internal format called a *coverage*. Along with the graphic information, the coverage stores a database of feature descriptions. A database is a file that contains columns (*items*) and rows (*records*) that hold tabular information about the graphic data. In the database there is one record for each graphic feature. For example, for each road on a map, there will be a corresponding record in the database describing that road.

Once the graphic data has been converted to a coverage and the database created, the information can then be grouped either by selecting an entity from the drawing or by specifying a value within the database. Groups of information can then be displayed in the AutoCAD drawing or printed in a report format.

A typical ArcCAD session uses AutoCAD to draft the graphic or spatial information. ArcCAD commands are then used to convert the graphics to a coverage. Database information can then be added to the graphics. The data is then available for query and analysis. Once analysis is complete, the results are redisplayed in AutoCAD.

INSIDE ArcCAD will expand on these GIS concepts. The remainder of this chapter outlines the components of an ArcCAD system.

The Components of an ArcCAD System

Release 11.4 is the AutoCAD Release 13 complement of ArcCAD. As an AutoCAD Development System (ADS) add-on to AutoCAD, ArcCAD has the same basic system requirements as AutoCAD.

Recommended Hardware

The recommended hardware configuration is a 90MHz Pentium processor or better with 32MB of RAM and a 1GB hard drive. A typical system also uses a 24"x36" digitizer and a 17-inch or larger high-resolution monitor.

Recommended Software

Microsoft Windows NT 3.5 is recommended, with a 65MB permanent swap file configured. The AutoCAD software should be Release 13 C3 or later.

The ArcCAD Configuration

By default, the ArcCAD properties should be configured to run the Arcadwin.exe file found in the c:\arcadsource\win directory where *arcadsource* is your install directory. A sample setup is shown in the following figure.

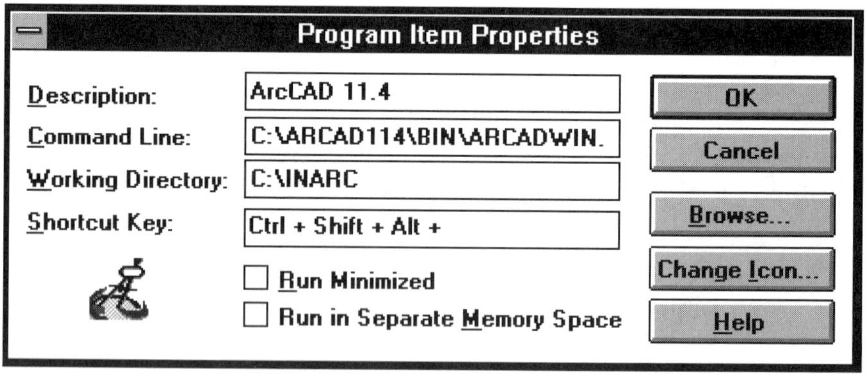

The Windows Properties dialog box.

Once the Properties dialog box has been set up or the batch file created, ArcCAD can be started.

From within the AutoCAD Drawing Editor, the ESRI_PREFS command found on the Tools menu is used to have ArcCAD automatically start whenever AutoCAD is loaded. Once the dialog box appears, select the Miscellaneous tab at the top.

The ESRI_PREFS dialog box.

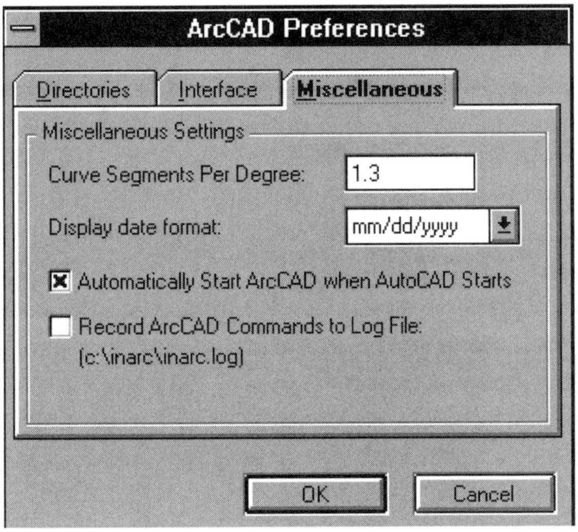

You should now see "Automatically start ArcCAD when AutoCAD starts." Toggle it on now and select OK.

AutoCAD's Role

AutoCAD's role in an ArcCAD GIS system is as a data input, manipulation, and display engine. Because ArcCAD stores spatial data in a coverage format, there is no need to store the graphic data in AutoCAD drawing files once the themes have been defined and coverages created. AutoCAD drawing files are used as source data for ArcCAD and to store map compositions, that is, maps that have been created by thematically displaying features in the ArcCAD coverage. These map compositions make use of all the features of AutoCAD, including layers, colors, line types, and hatch patterns. Coverages, themes, and thematic mapping will be covered in depth in other chapters.

Command Input

ArcCAD commands are available through all of the typical AutoCAD input methods, including side screen and pulldown menus, command prompt input, and the digitizing tablet.

Customizing Command Input

ArcCAD command aliases may be defined for commonly used commands with the ACAD.PGP file. For automating repetitive tasks, ArcCAD commands may also be used within AutoCAD scripts and AutoLISP and ADS functions.

The ArcCAD Interface

In the next chapter you will work through a session explaining the basic concepts of ArcCAD. When you start ArcCAD, you will notice that the regular AutoCAD pulldowns have been replaced by an ArcCAD menu. You can easily execute AutoCAD commands by selecting them from the sidescreen menu or by typing them at the command prompt.

The ArcCAD menu.

ArcCAD makes extensive use of AutoCAD's dialog boxes. Most ArcCAD commands can be run from within a dialog box. These commands use the syntax of "DD" in front of the standard command names. For example, the DDTheme command provides all of the options available from the Theme pulldown menu.

The Theme menu.

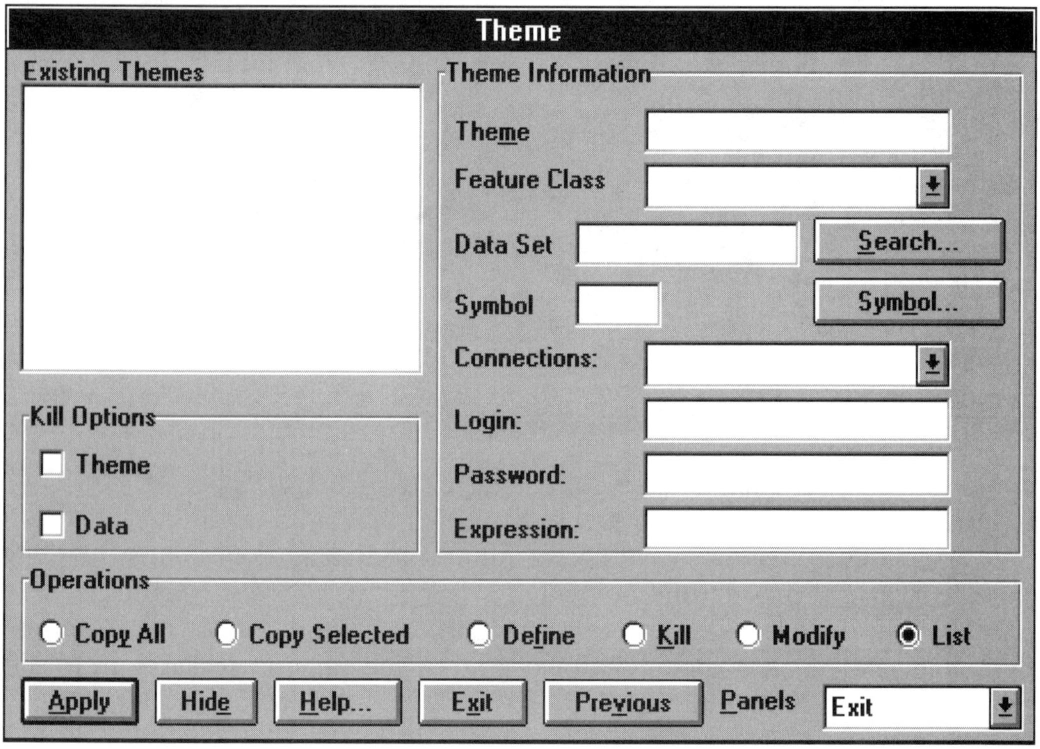

The Theme dialog box.

The ArcCAD Database Screen

ArcCAD's one addition to the AutoCAD interface is the database screen. This screen is a DOS program that can be used for displaying and editing items and records from a database. The database screen, as shown in the following figure, has a row of command buttons along the top. These commands allow you to scroll through the data, moving up or down an item or a page at a time, directly to a specific record, and choosing which items and records to display.

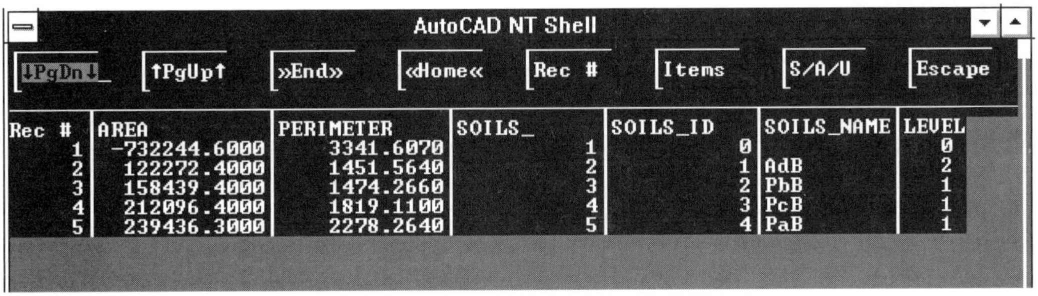

The ArcCAD Database screen.

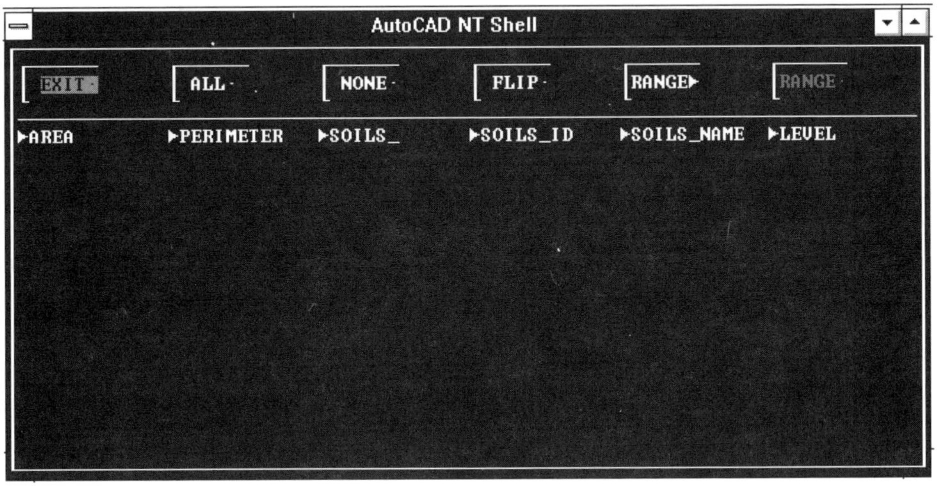

The ArcCAD Database screen showing Item control.

The ESRI_PREFS command on the Tools pulldown menu controls whether or not this DOS-based database screen is used. If the DOS-based editor is turned off, either the AutoCAD text screen will be used to view and edit all databases in ArcCAD or the Wbrowse command must be used to view and add databases in a dialog box.

For more information on the ArcCAD menu, navigating dialog boxes, and the ArcCAD database screen, refer to Chapter 2.

Review

In this chapter you have been introduced to GIS and the ESRI product line. You have been given a brief introduction to ArcCAD: the components of ArcCAD, how ArcCAD works, including the database and items screens, and how to interface with these components. You should also have an understanding of AutoCAD's role within the ArcCAD environment.

Chapter

Sample Work Session

Putting ArcCAD To Work

This work session is intended to be a walkthrough of ArcCAD. It will introduce you to the menu structure, commands, and processes involved in a typical session of ArcCAD. During this session, you will use commands and see terms that you may not be familiar with. Don't worry; these commands and terms will be discussed in depth as you move through the book.

In this exercise you will create a GIS data set from an AutoCAD drawing, analyze the data set, and then map the data set based on several database properties.

Before beginning the lesson, be sure that the training data set is installed. For installation help, see "Installing the Data Disk."

Setup

To begin the lesson, start the ArcCAD system.

From Windows, double-click on the ArcCAD icon.

or

From DOS, type the batch file name at the C:\ prompt and press Return.

Once in the drawing editor, open the INARC1 drawing. If ArcCAD does not start automatically, start it now. For help on starting ArcCAD within the drawing editor, refer to the *ArcCAD User's Guide.*

The drawing is shown in the following figure.

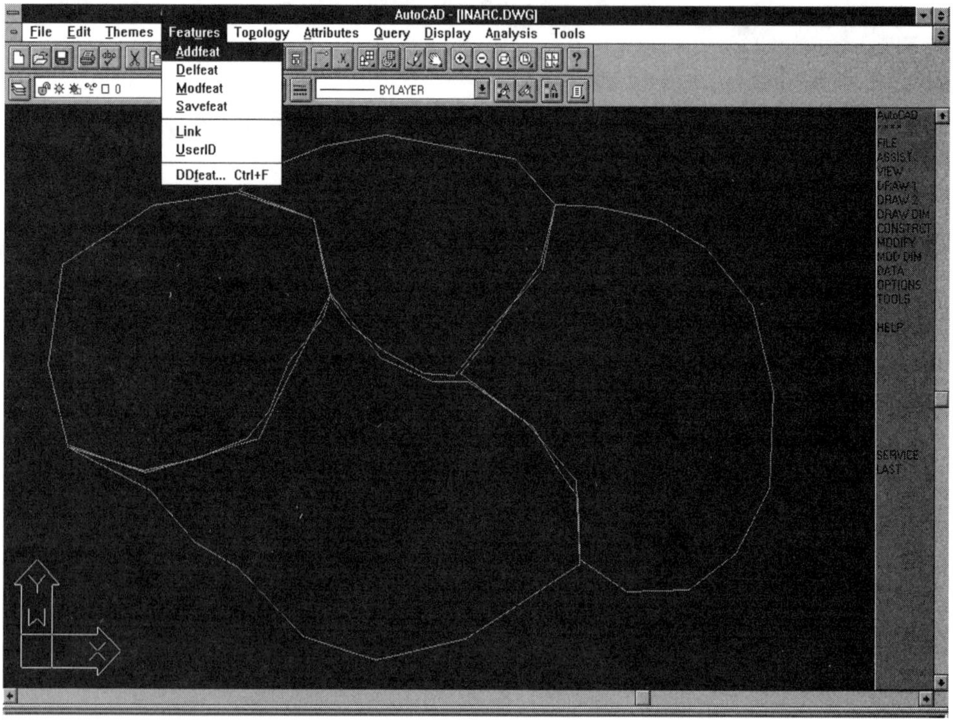

The INARC1 drawing file.

The INARC1 drawing file contains a soils map of an undeveloped parcel of land. In this exercise you will create a GIS dataset from this soils map, add information to the database, and display this dataset back into AutoCAD.

Once ArcCAD is loaded, the ArcCAD menu will appear.

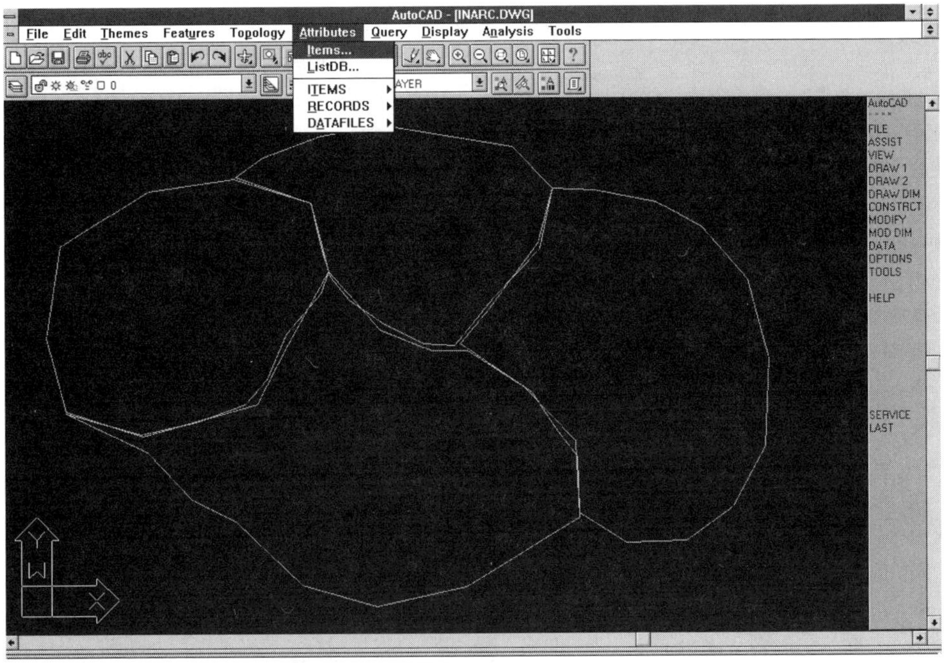

The ArcCAD menu.

The ArcCAD menu is broken down into three distinct sections. The first section includes the Themes, Features, and Topology menus. These three menus make up the spatial data creation and editing portion of ArcCAD.

The second section consists of the Attributes menu. All database or feature attribute table (FAT) creation and editing commands can be found on this menu.

The final section of the menu consists of the Query, Display, and Analysis menus. These three menus make up the analysis and display portion of ArcCAD.

Creating a Dataset

Before you begin creating coverages (ArcCAD's internal format for storing data), you must first tell AutoCAD the coverage name and where it will be stored. This is done by defining a theme. A *theme* is ArcCAD's method of accessing coverage data. The theme stores the coverage location and type.

To define a theme, select Defthm from the Themes pulldown menu

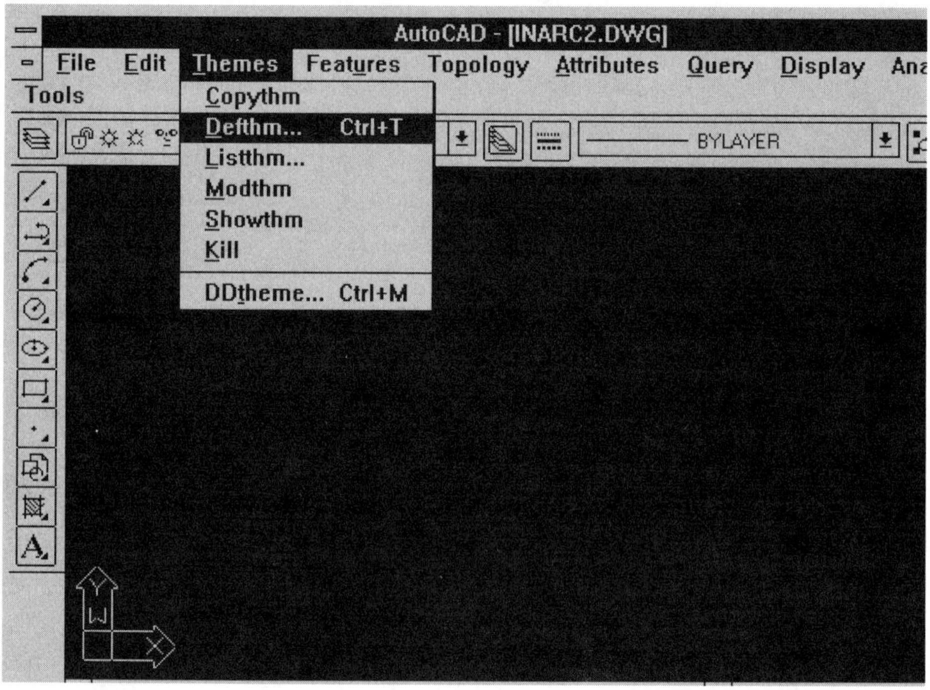

The Defthm menu.

Fill in the Defthm dialog box as shown in the following figure.

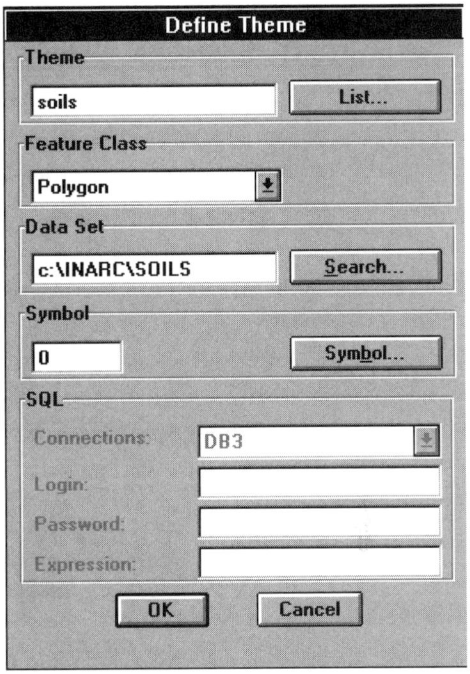

The Defthm dialog box.

Once the dialog box has been filled in, select OK to finish the command. The next step is to add data by using the Addfeat command from the Features menu.

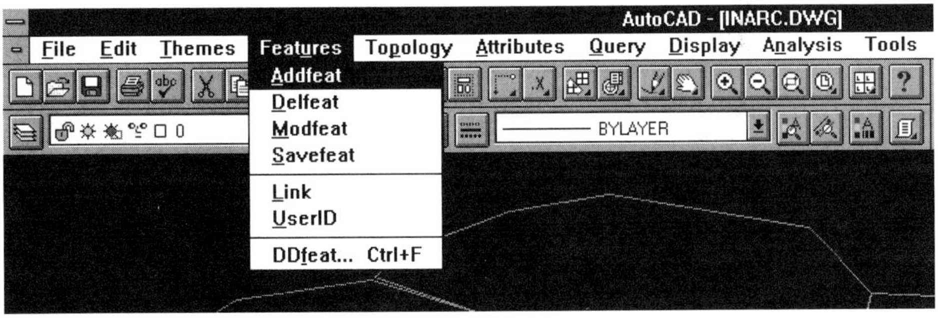

The Features menu.

The Addfeat command will add a representation of each AutoCAD entity selected in the spatial database and create a FAT for the theme.

Follow the input below.

```
Command: addfeat
Theme name (?/<SOILS>): ENTER
Select objects: 220,130
Other corner: 1650,1000
4 found
Select objects: ENTER
Optional property table (?/<none>): ENTER
4 written to C:\inarc\SOILS, 0 duplicated, 0 ignored.
```

ArcCAD now displays a dialog box, prompting whether to Clean the dataset, Build the dataset, or Wait and do neither.

The Clean/Build dialog box.

Select Clean and follow the input below.

```
Output theme (?/<SOILS>): ENTER
Dangle length <default>): 20
Fuzzy tolerance <compute>: 20
Cleaning SOILS.
Sorting...
Intersecting...
Assembling polygons...
  Sorting input file...
  Sorting label file...
  Processing...
  Assigning final IDs...
  Writing arc file...
  Generating polygon report...
Creating Polygon attribute table.
Dropping entity-feature links...
```

You have just created your first spatial dataset. ArcCAD has stored the four soils polygons from the INARC1 drawing file to a coverage whose location and name are defined within the SOILS theme.

Let's take a look at the FAT that was created for the SOILS theme.

To view the FAT, select Wbrowse from the Attributes menu under DATAFILES.

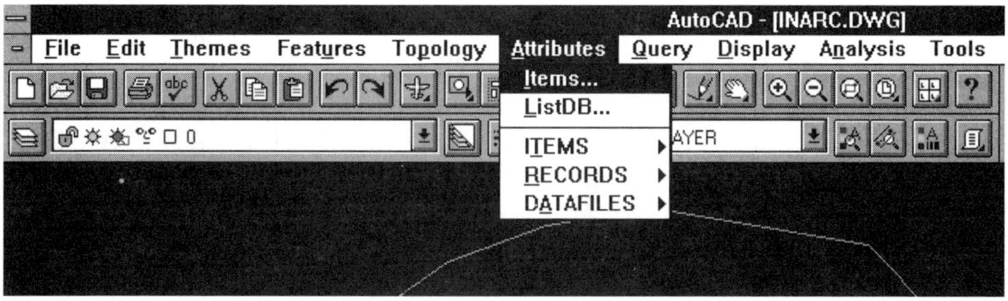

The Attributes menu.

At the theme name prompt, press Enter to accept the default of SOILS. The FAT is shown in the following figure.

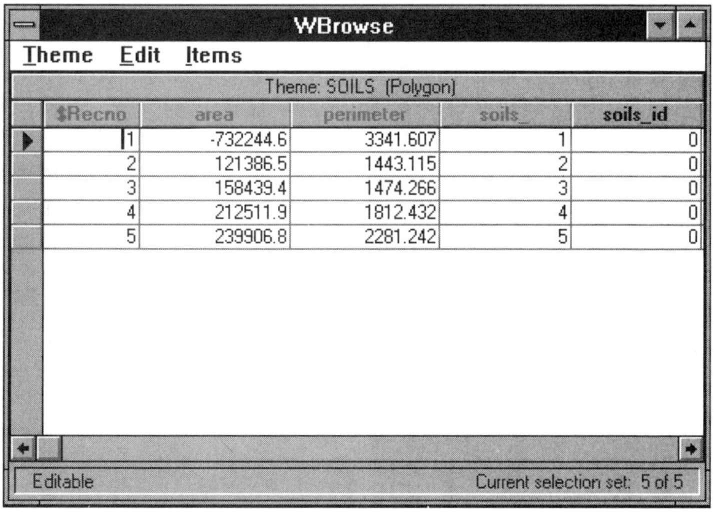

The SOILS FAT.

The FAT shows the four fields automatically created for a polygon theme. These four fields are area, perimeter, an internal ID, and a User-ID. The two IDs created by ArcCAD are important to ESRI's spatial data model. In the case of the SOILS coverage, the two IDs are named SOILS_ and SOILS_ID. The internal ID, denoted by the coverage name followed by an underscore (_), is used by ArcCAD to relate the feature data to the graphic definition of the polygon. The User-ID, denoted by the coverage name followed by an "_ID" is used by the ArcCAD operator to identify which record in the FAT relates to each polygon in the spatial database.

Creating Labels and IDs

At this point values have not been assigned to the User-ID fields. This is due to the lack of label points within the polygons. The concept of polygon label points will be discussed later. For now you will learn how to have ArcCAD create labels and assign IDs for you.

Exit the Wbrowse command by selecting Exit from the dialog's Theme menu.

To create labels and IDs for the soil polygons, select Createlab from the Topology menu.

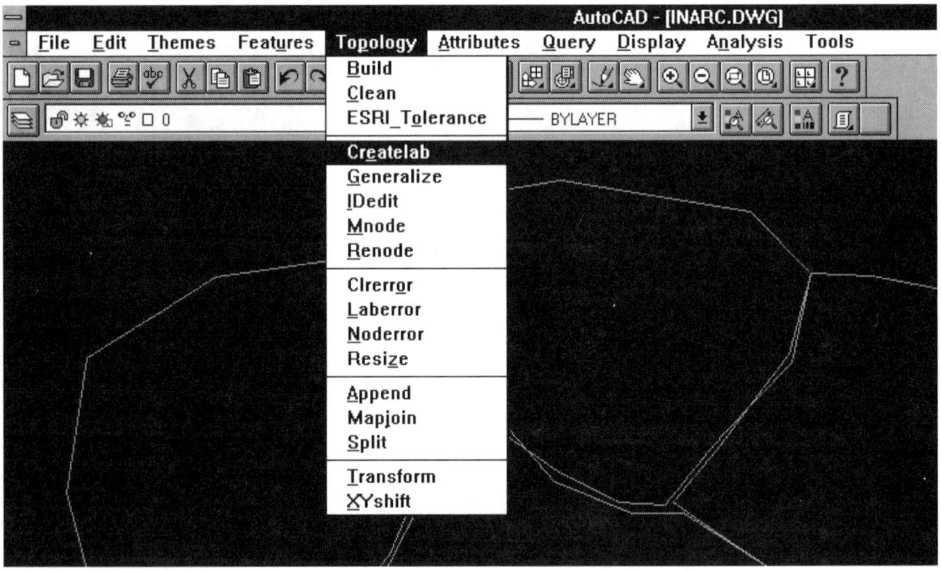

The Topology menu.

At the theme name prompt, press ENTER to accept the default of SOILS and ENTER again at the User-ID base.

```
Command: createlab
Theme name (?/<SOILS>): ENTER
User-ID base <0>: ENTER
```

ArcCAD has now created a temporary file holding User-IDs for the soils polygons. To add the IDs to the polygon coverage, the coverage must first be built to update the topology.

From the Topology menu choose Build. Press ENTER at the *Input theme,* and *Output theme* prompts to accept the default of SOILS.

```
Command: BUILD
Input theme (?/<SOILS>): ENTER
Output theme (?/<SOILS>): ENTER
```

ArcCAD now presents a dialog box, warning you that specifying the same output theme name as the input name will destroy links to entities in the drawing. Since you have not yet displayed ArcCAD coverage information into the drawing, there are no links to destroy.

Press Yes to continue building the coverage.

This completes the process of adding label points and User-IDs to the SOILS polygon coverage.

To see the updated FAT, select Wbrowse from the Attributes DATAFILES menu. The new FAT should look like the following figure.

The SOILS FAT with IDs added.

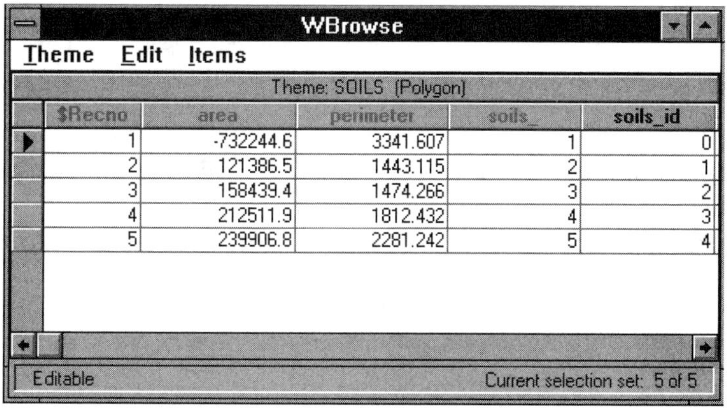

$Recno	area	perimeter	soils_	soils_id
1	-732244.6	3341.607	1	0
2	121386.5	1443.115	2	1
3	158439.4	1474.266	3	2
4	212511.9	1812.432	4	3
5	239906.8	2281.242	5	4

Select Exit from the Dialog's theme menu to return to the drawing editor.

The next step in the creation of this dataset is to add the names of the soils to the FAT. You first need to know which polygon is related to which record in the FAT. To identify this relationship you will display the polygons into the drawing file and display the User-IDs within the polygons.

Before bringing the coverage polygons into the drawing and labeling them, you will need to create a new working layer and set the text size for the labels.

```
Command: Layer
?/Make/Set/New/ON/OFF/Color/Ltype/Freeze/Thaw/
  LOck/Unlock: m

New current layer <SOILS>: work
?/Make/Set/New/ON/OFF/Color/Ltype/Freeze/Thaw/LOck/
  Unlock: of
Layer name(s) to turn Off: *
Really want layer WORK (the CURRENT layer) off?
?/Make/Set/New/ON/OFF/Color/Ltype/Freeze/Thaw/LOck/
  Unlock: ENTER
Command: TEXTSIZE
New value for TEXTSIZE : 20
```

With the current layer set to WORK, select Dropline from the Display, LINES menu.

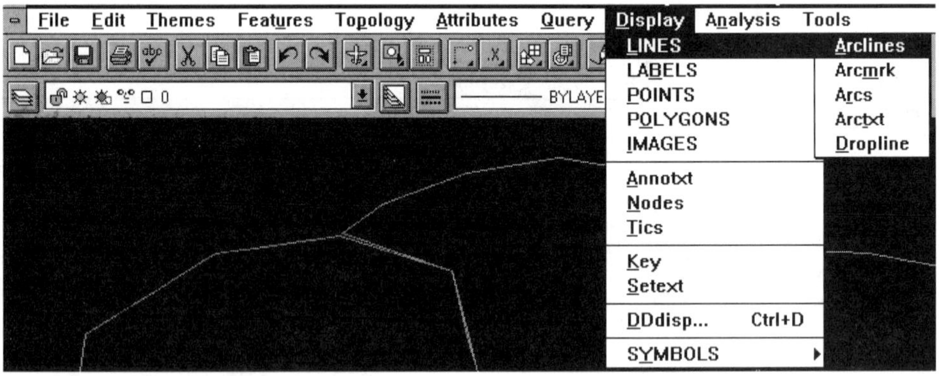

The Display LINES menu.

```
Command: DROPLINE
Theme name (?/<SOILS>):
Item (?): ?
```

Entering a ? at the *Item* prompt displays an Items Selection dialog box. This dialog box presents a list of all items in the selected theme, with definitions and values for each item.

The Item Selection dialog box.

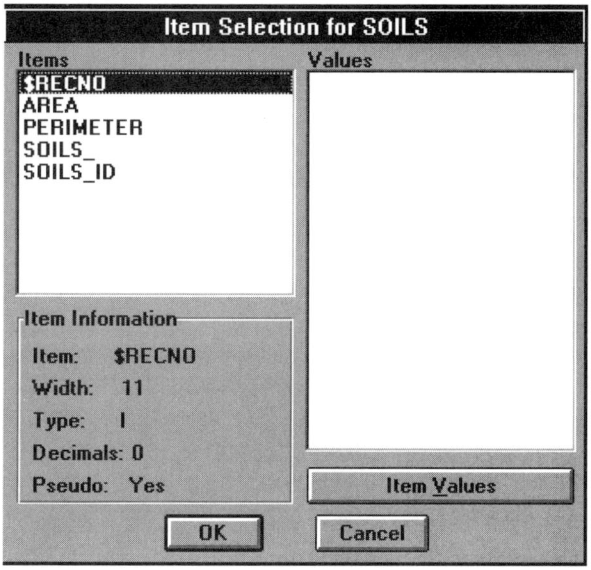

Select SOILS_ID from the list and press OK.

```
Optional Lookup table (?/<none>): ENTER
Line symbol number (0-100) (?/<0>): ?
```

Entering a ? at the *Line symbol number* prompt brings up a dialog box, allowing the graphical selection of a line symbol number.

The LINE Symbols dialog box.

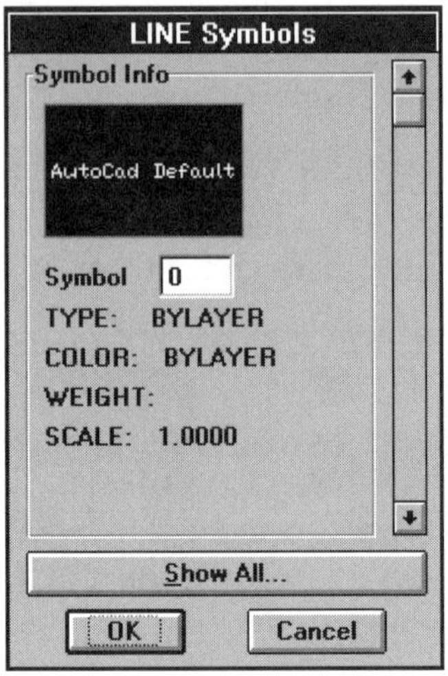

To display all symbols available, choose Show All from the dialog box.

The LINE Symbol dialog box.

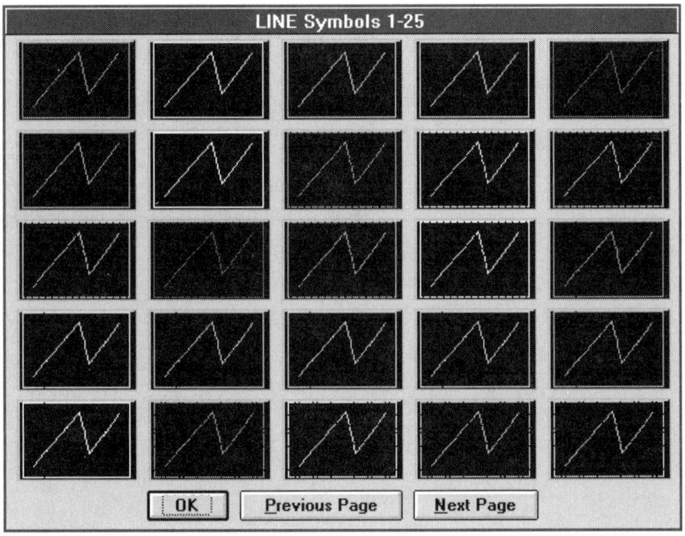

Choose the fourth symbol in the top row, cyan, and then choose OK to exit.

```
Label polygons? (No/<Yes>): ENTER
```

Answering Yes at the *Label polygons* prompt will cause ArcCAD to label each polygon with the value of the item selected earlier.

The *Text symbol number* prompt then allows for the selection of a text symbol.

For now accept the default of 0.

```
Text symbol number (0-100) (?/<0>): ENTER
```

The SOILS polygons are now drawn on layer WORK with a color of cyan.

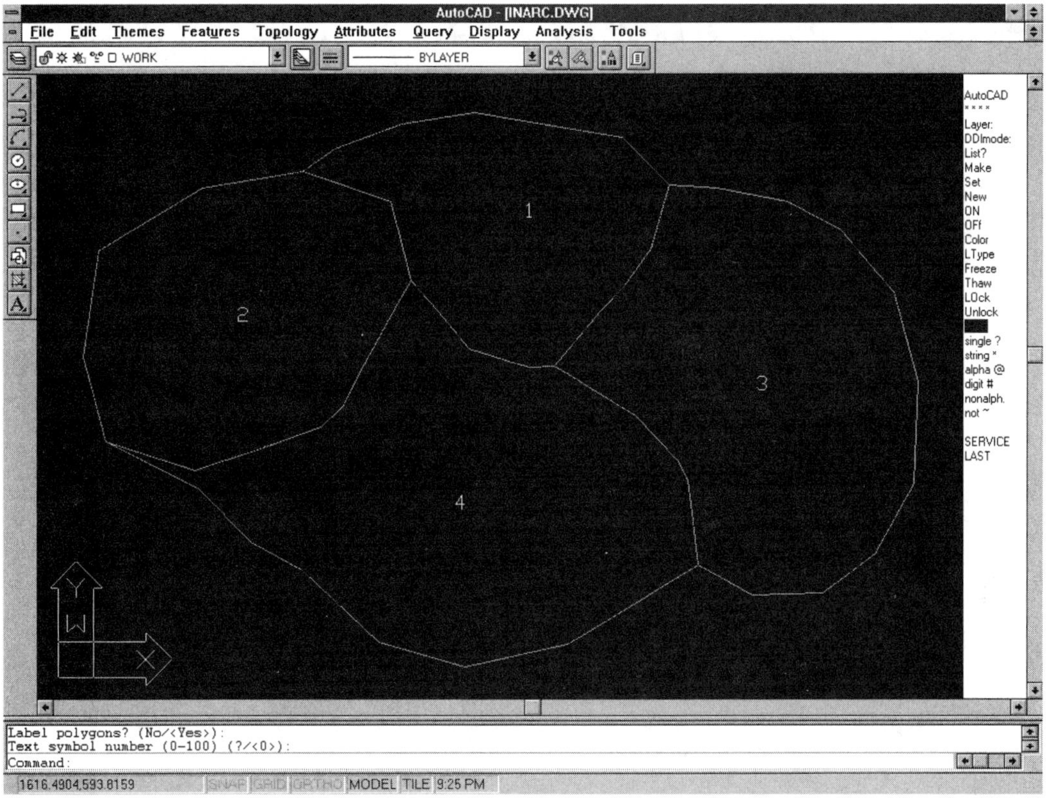

The Soils Polygons from the coverage.

You will notice that the polygons displayed from the coverage look slightly different than those from the original drawing. This is a result of the Clean command. The Clean command has snapped all common lines together into single lines where the separation was less than the 20-foot fuzzy tolerance you specified in the Addfeat command.

With the polygons and their User-IDs now displayed in the drawing, quick inspection shows you that the top center polygon is attached to the record, with User-ID equal to 1. The rightmost polygon is attached to the record, with User-ID equal to 3 and so on. With this information you are now ready to add the soil name information to the FAT.

Adding Soil Information to the Feature Attribute Table

Adding information to the FAT is a two-step process. First, an Item or Column must be added to the table. Second, the item values, or row information, must be added. You will now add an item to the soils FAT called SOILS_NAME. You will then modify the four polygon records to add the name of each soil.

From the Attributes menu, choose ITEMS and then Additem.

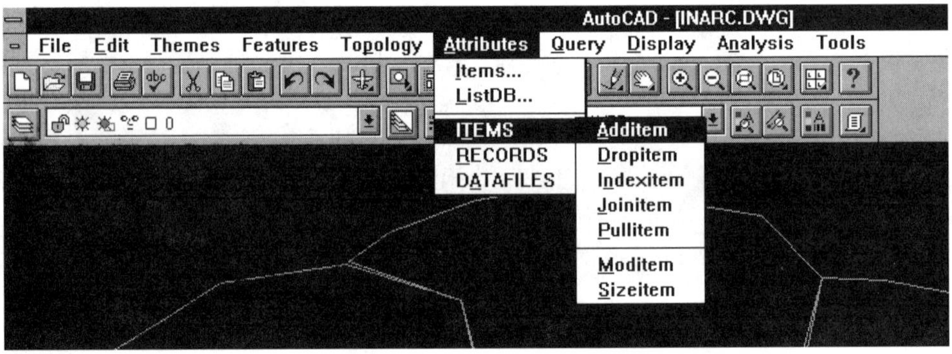

The Additem menu.

At the *Theme name* prompt press Enter to accept the default of SOILS. At the *Item name* prompt enter SOILS_NAME and follow the input below.

```
Command: additem
Theme name (?/<SOILS>): ENTER
Item name (?): SOILS_NAME
Item width: 5
```

```
Item type?  (Character/Date/<Numeric>): C
Start item <last>: ENTER
```

The Additem command has added a new column to the FAT called SOILS_NAME.

The Wbrowse command found on the Attributes menu under Datafiles will be used to add the soil names to the new column in the FAT. This command will display the database in a spreadsheet-like dialog box for editing.

```
Command: WBROWSE
Theme name (?/<.>): SOILS
```

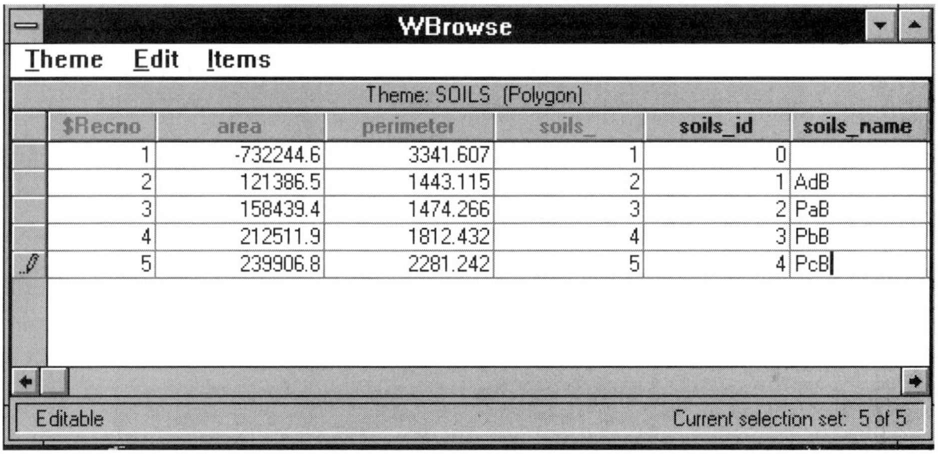

$Recno	area	perimeter	soils_	soils_id	soils_name
1	-732244.6	3341.607	1	0	
2	121386.5	1443.115	2	1	AdB
3	158439.4	1474.266	3	2	PaB
4	212511.9	1812.432	4	3	PbB
5	239906.8	2281.242	5	4	PcB

The Wbrowse dialog box for SOILS.

To fill in the soil names, pick a point in the respective cell and type in the name of the soil. Be sure to press Enter after each entry to save the value. Once all of the soil names have been entered, select exit from the dialog box's Theme menu.

To verify that all of the polygons now have a soil name and that each soil name is attached to the correct polygon, use the Identify command found on the Query menu.

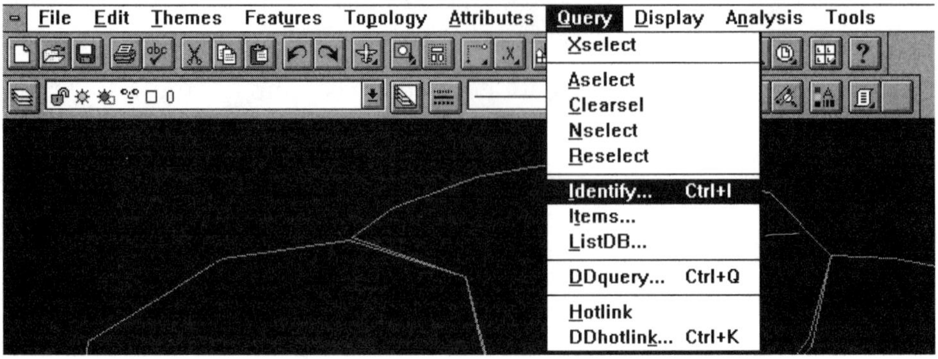

The Query menu.

At the *Theme name* prompt press ENTER to accept the default of SOILS. The Identify command now prompts you to select a polygon. Simply pick a point within any of the polygons.

```
Command: identify
Theme name (?/<SOILS>): ENTER
Select a polygon:
```

After selecting a polygon, ArcCAD displays the appropriate record within a dialog box.

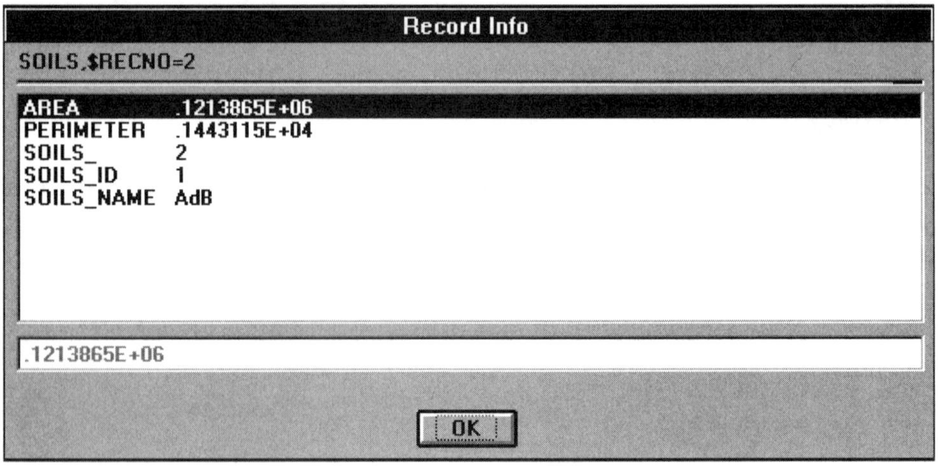

The Record Info dialog box.

After reviewing the record information for the selected polygon, press OK from this dialog box and select another polygon to display. This allows the verification of data entry for the soil polygons. In later chapters you will learn a similar method of adding information to features.

With the verification process complete, you have created your first GIS dataset. The dataset has been cleaned, User-IDs have been created, and user-defined attribute data has been added. You can now use this data to create thematic maps of the soils based on information contained in the FAT.

Creating a Thematic Map

In the next section of this work session you will create a thematic map using a lookup table that shows each soil polygon with shading based on its soil name. A lookup table is a database that is used to associate feature attributes with symbol numbers or labels.

Once the shading is complete you will label each polygon with its soil name and store the map as a new AutoCAD drawing.

To begin, you must define a theme identifying the name and location of the lookup table you will use to shade the soil polygons. You used the Defthm command at the beginning of this chapter to define a polygon theme. You will now use the Defthm command to define a record theme. Choose Defthm from the Themes menu and fill out the dialog box as shown below.

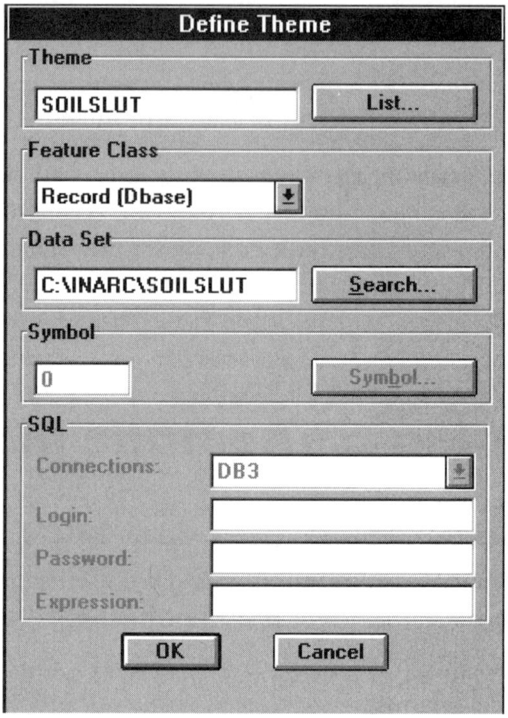

The Defthm dialog box for the SOILS lookup table.

The lookup table database already exists. Defining a theme for it makes it accessible.

To view the contents of the Soils lookup table, choose Wbrowse from the Attributes Datafiles menu.

At the *Theme name* prompt press ENTER to accept the default of SOILSLUT.

Notice that the default theme name has changed from SOILS to SOILSLUT. ArcCAD always uses the last theme defined or accessed as the default. Since SOILSLUT was just defined, it has become the default. The table should look like the one shown in the following figure.

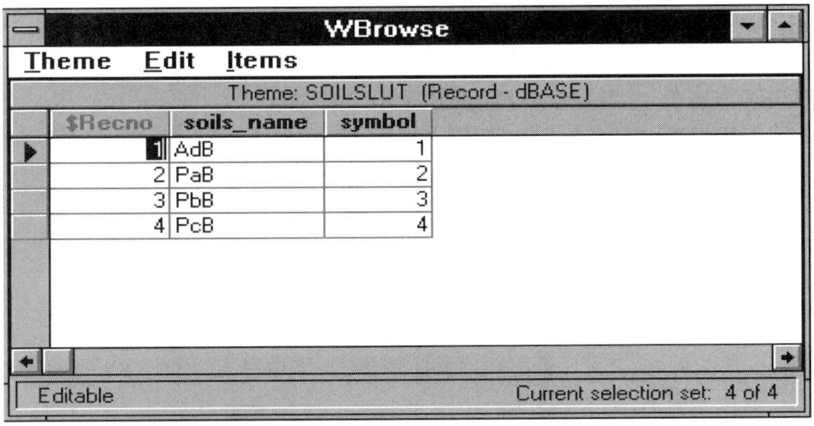

The Soils lookup table.

The lookup table contains two items. The first is the SOILS_NAME item from the SOILS coverage. Notice that the soil names are listed in alphabetical order within this table. The second is the SYMBOL number with which to display each polygon. ArcCAD will match a SOILS_NAME in the SOILS coverage with the SOILS_NAME in the lookup table and display the polygon with the corresponding symbol number.

To return to the drawing editor, select Exit from the dialog box's Theme menu.

Before displaying the new information within the drawing, let's clear the screen of the previous polygon display.

Use the AutoCAD ERASE command and window the entire drawing.

Now, to display the polygons, select Polyshd from the Display, Polygons menu.

At the *Theme name* prompt, type SOILS and follow the input below.

```
Command: polyshd
Theme name (?/<soilslut>): soils
Display using? (Properties/Item/<Symbol>): i
Item (?): soils_name
Optional lookup table (?/<none>): soilslut
```

You should now see the four soils polygons displayed in red, yellow, green, and cyan. The last step in creating your thematic map is to add the soil names to the display. This is accomplished by the Polytxt command found on the Display, Polygons menu.

Select Polytxt and follow the input below.

```
Command: polytxt
Theme name (?/<SOILS>): ENTER
Item (?): ?
```

The Item Selection dialog box.

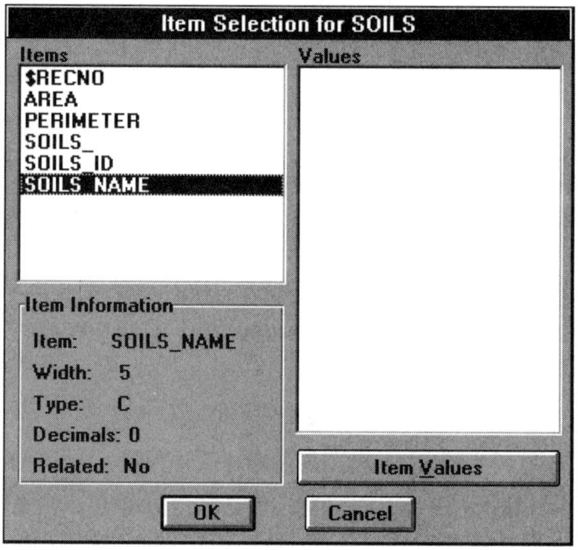

Choose SOILS_NAME from the Item Selection dialog box.

```
Optional lookup table (?/<none>): ENTER
Text symbol number (0-100) (?/<0>): 5
```

With the polygons labeled with their soil name, the completed soils map should look like the following map.

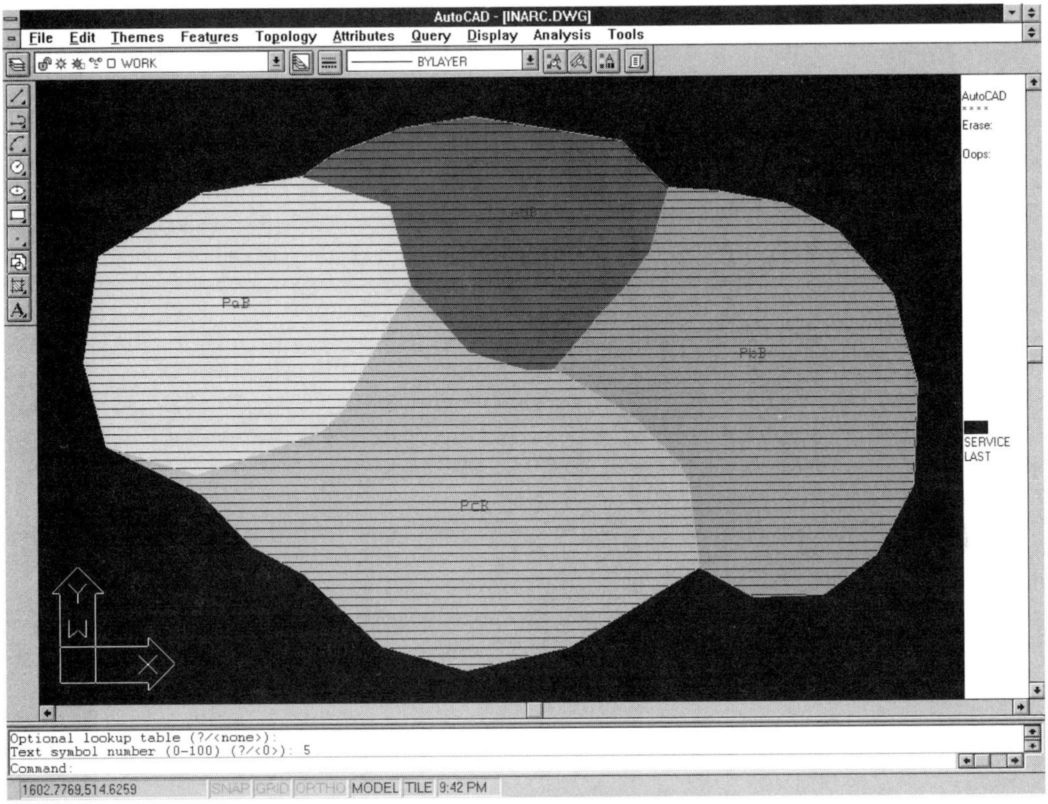

The completed soils thematic map.

Adding a Legend

One addition you will typically want to make to a thematic map is to add a legend, rather than placing labels on top of the shading. To create a legend for your soils map, you will first create a key file and then display the key file into the drawing. Let's take a look at this process now.

To create a key file you will use a simple text editor.

If you are running in Windows use the ALT TAB key combination to return to the Program Manager. Once there, open the Accessories group and execute Notepad. If you are running in DOS, type SHELL at the command prompt and press ENTER twice. Once at the DOS prompt type "EDIT SOILS.KEY."

Key files in ArcCAD are very simple. Each group of two lines makes up a definition. The first line contains a decimal point and a symbol number. The next line contains the label for that symbol.

Enter the information shown below into your key file.

The Soils map with a key being created.

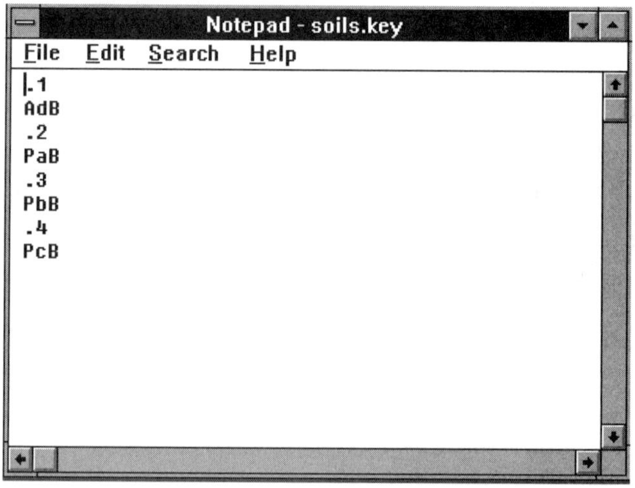

Once the information has been entered as shown previously, save the key file as \INARC\SOILS.KEY. If you are running in Windows, use Alt + Tab to switch back to the ArcCAD session. If in DOS type EXIT to return to the AutoCAD drawing editor.

The key file can now be used to draw a legend in the drawing using the Key command found on the Display menu.

To create the key, follow the input below.

```
Command:KEY
KEY Symbol class? (Line/Marker/Shade): S
Key file name: C:\INARC\SOILS.KEY
Width of key box <0.7500>: 150
Height of key box <0.7500>: 100
```

```
Horizontal separation <0.2500>: 25
Vertical separation <0.2500>: 25
Draw key box outline? (No/<Yes>): ENTER
Key upper left corner (X,Y):1700,800
```

The new drawing should look like the one shown in the following figure.

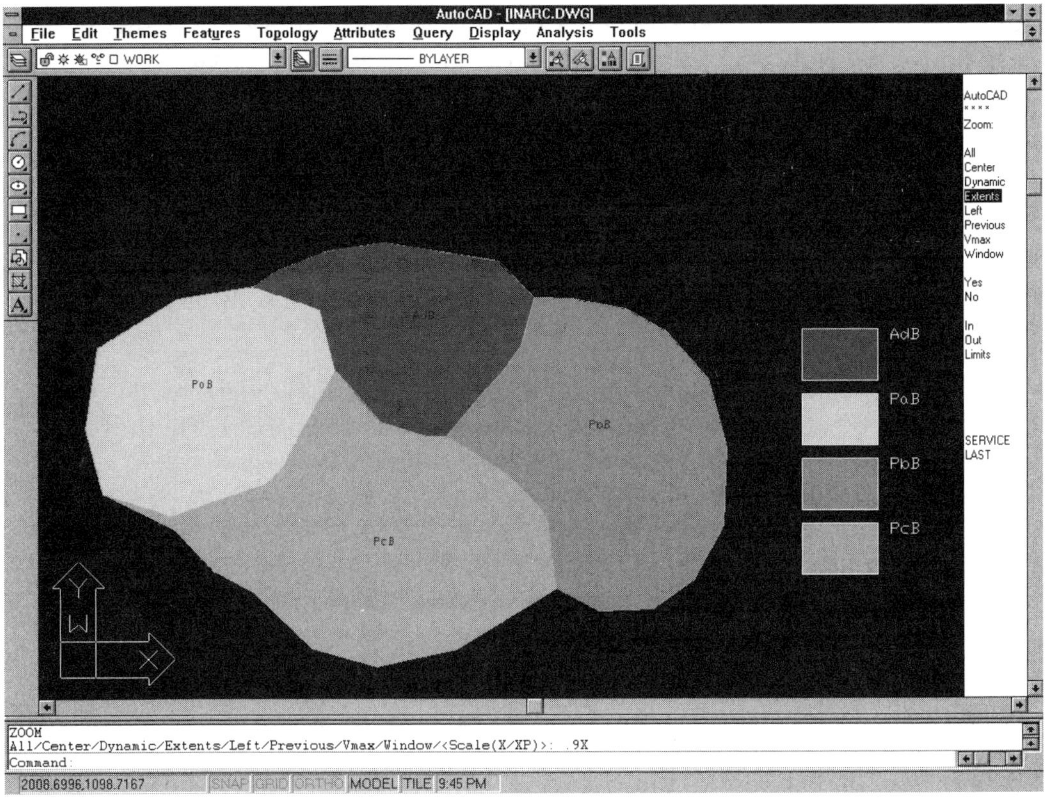

The Soils map with a key.

Analyzing Data

So far you have been introduced to many of ArcCAD's abilities. The one area we have not touched on is ArcCAD's analysis capabilities. You will work through a quick example here.

The soils map you produced shows each soil in a different color. Although this is typical, a soils map may also be used to indicate drainage. Each soil is assigned a drainage level, and based on that level, the soils are aggregated. This process groups multiple soil polygons with the same drainage level into a single polygon, removing common boundaries. In your dataset, soils with a type of PaB, PbB, and PcB are assumed to have a drainage level of "1." Soil type AdB has a drainage level of "2." Using these drainage levels you will aggregate the soils.

Before beginning this process you must first add the drainage information to your soils FAT. In this example a separate database has already been created with the drainage level information. This database was created separately because it can be used throughout many projects. In all projects each soil name always has the same drainage level. Since the drainage level database already exists, adding this information to the soils FAT will be handled slightly differently.

> Start by creating a theme that accesses the DRAINLVl database. The theme is created using the Defthm command found on the Themes menu.
>
> Use DRAINLVL as the theme name, RECORD as the theme type, and C:\INARC\DRAINLVL as the dataset name.
>
> Once the dialog box has been completed, select OK.

When this theme is accessed it will automatically check to see if a database with the path and name specified already exists. If it finds a database it will use that information to carry out the specified command.

> To see the DRAINLVl database, use the Wbrowse command found on the Attributes menu under DATAFILES. The database is shown in the following figure.
>
> To return to the drawing editor select Exit from the Dialogs Theme menu.
>
> To permanently add the items and records in the DRAINLVL database to the information in the Soils FAT you will use the Joinitem command found on the Attributes, Records menu.

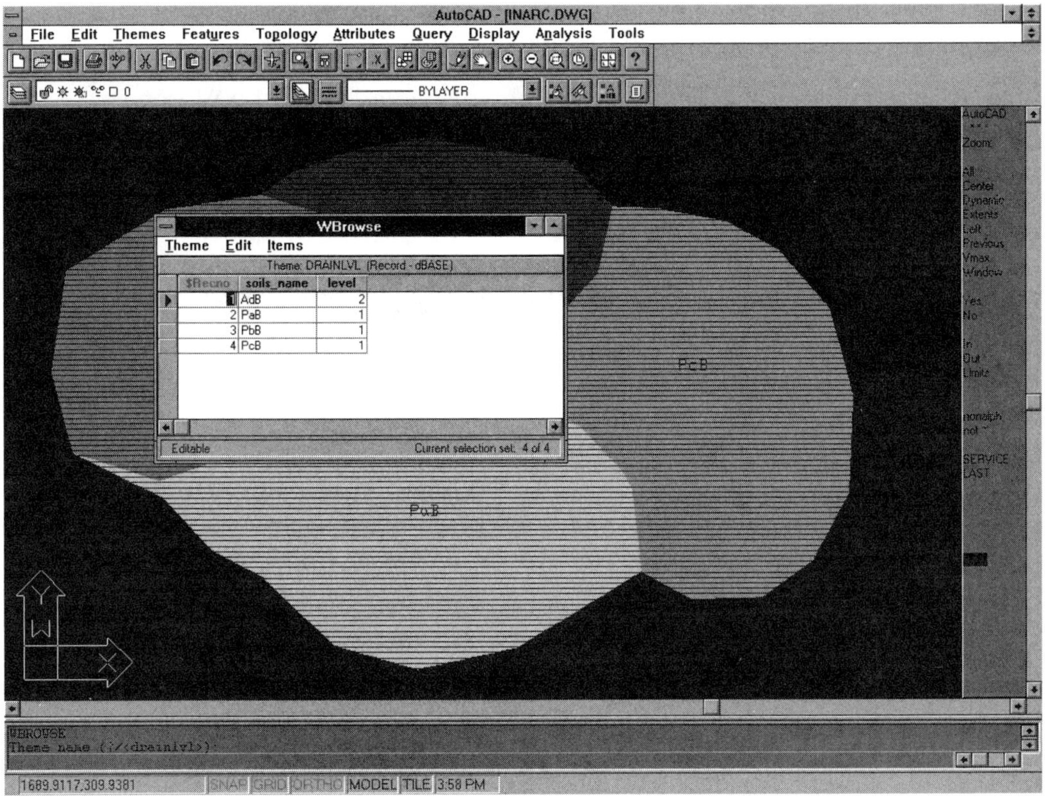

The DRAINLVL database.

To complete this process, follow the input below.

```
Command: JOINITEM
Input theme (?/<DRAINLVL>): SOILS
Join theme (?): DRAINLVL
Output theme (?/<SOILS>): ENTER
Relate item (?): SOILS_NAME
Start item (?/<last>): ENTER
Join method? (Link/Ordered/<LInear>): ENTER
Item "SOILS_NAME" duplicated, Join File version
  dropped
```

The last message returned by the Joinitem command is that Item SOILS_NAME was duplicated and dropped. This is because the item SOILS_NAME existed in both the SOILS FAT and the DRAINLVL database. The SOILS_ NAME in the Input theme was left intact.

The combined database may now be displayed using the Wbrowse command from the Attributes DATAFILES menu. At the *Theme name* prompt press ENTER to accept the default of SOILS.

The joined tables are shown in the following figure.

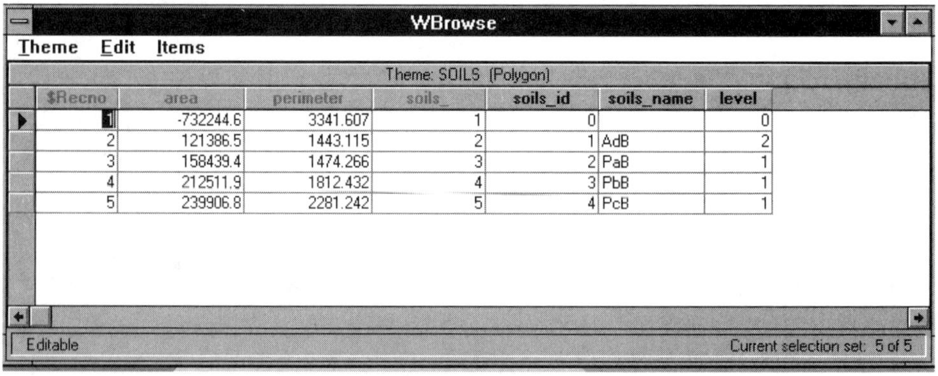

$Recno	area	perimeter	soils_	soils_id	soils_name	level
1	-732244.6	3341.607	1	0		0
2	121386.5	1443.115	2	1	AdB	2
3	158439.4	1474.266	3	2	PaB	1
4	212511.9	1812.432	4	3	PbB	1
5	239906.8	2281.242	5	4	PcB	1

The joined SOILS and DRAINLVL tables.

Combining the two tables provides you with the necessary information to aggregate the soils. This is done with the Dissolve command found on the Analysis menu.

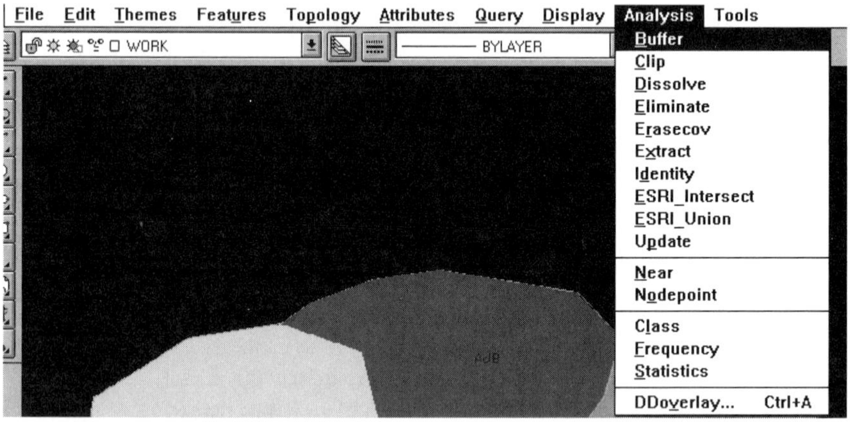

The Analysis menu.

To aggregate the soils, follow the input below.

```
Command: Dissolve
Input theme (?/<SOILS2>): SOILS
Output theme (?): SOILSLVL
Theme SOILS2 is not defined.
Defining SOILS2...
First item (?/<#ALL>): LEVEL
Last item (?/<LEVEL>): ENTER
Dissolving SOILS by LEVEL+LEVEL to create SOILS2.
Creating C:\INARC\SOILS2.PAT format...
Creating dissolve table...
Dissolving...
Number of Polygons (Input,Output) =        5         3
Number of Arcs     (Input,Output) =        9         3
Creating polygon attribute table...
```

To see the results of the Dissolve command, let's clear the screen and display the new polygons from the SOILSLVL theme.

You can clear the screen by either erasing all entities or by using the AutoCAD WBLOCK command to write the current map display to a new drawing for later use. At this point let's use the WBLOCK command and create a new drawing named INARCSLS. For more information on the WBLOCK command, refer to the *AutoCAD Reference Guide*. Your screen should now be clear.

The procedure you will use to create the Drainage Level map is identical to that used to create the soils map. Both the drainage level lookup table and the drainage level key file have been provided for you.

To create the drainage level map, define the DRAINLUT theme with the Defthm command found on the Themes menu.

The theme name should be DRAINLUT, the type is RECORD, and the dataset name is C:\INARC\DRAINLUT.

Now follow the input below to complete the map.

```
Command: POLYSHD
Theme name (?/<SOILSLVL>): ENTER
Display using? (Properties/Item/<Symbol>): I
Item (?): level
Optional lookup table (?/<none>): drainlut
```

Polyshd displays all of the polygons as solid fill. A lookup table named DRAINLUT was used to set the color for each polygon. You may view the DRAINLUT lookup table using the Wbrowse command.

Now add the key to the drawing file.

```
Command: KEY
KEY Symbol class? (Line/Marker/Shade): S
Key file name: C:\INARC\DRAINLVL.KEY
Width of key box <150.0000>: ENTER
Height of key box <100.0000>: ENTER
Horizontal separation <25.0000>: ENTER
Vertical separation <25.0000>: ENTER
Draw key box outline? (No/<Yes>): ENTER
Key upper left corner (X,Y): 1700,800
```

With the polygons shaded on screen you can outline them with a darker color by using the Polys command found on the Display, Polygons menu.

```
Command: POLYS
Theme name (?/<SOILSLVL>): ENTER
Line symbol number (0-100) (?/<0>): 5
```

The completed drainage map is shown in the following figure.

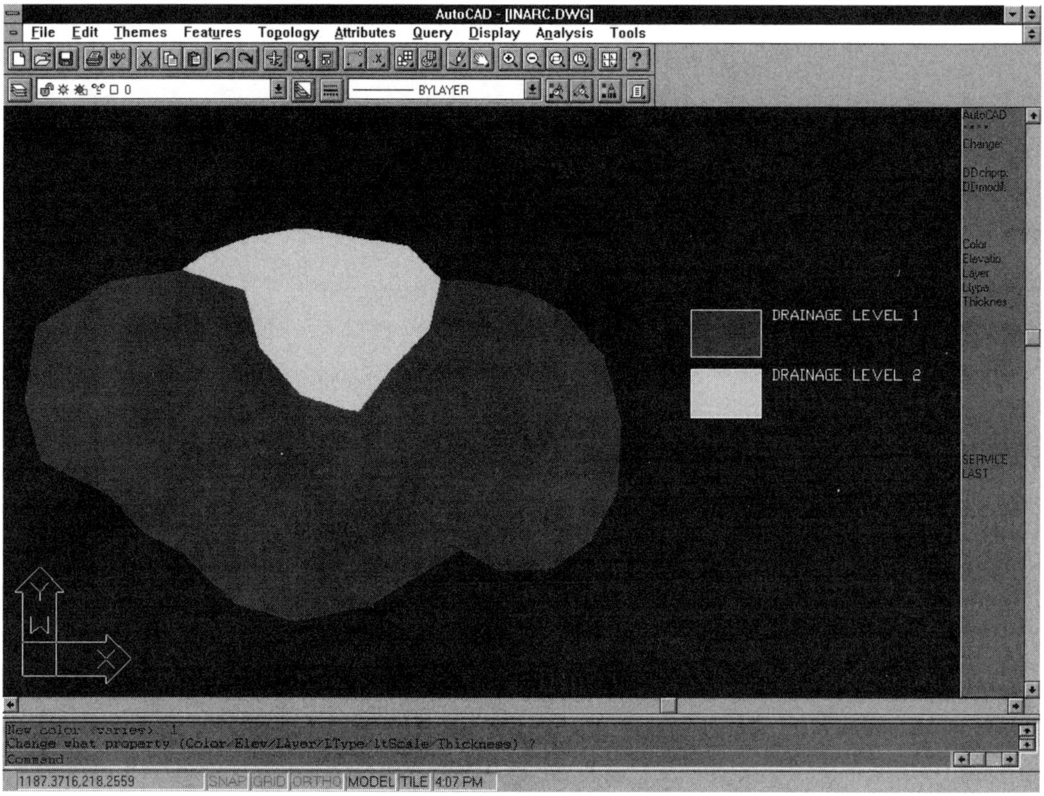

The completed drainage level map.

Review

You have now created two GIS datasets and two thematic maps. You have used lookup tables and key files and you have done spatial analysis. You should be comfortable with AutoCAD's role within an ArcCAD-based GIS system. All this and you haven't reached Chapter 3 yet! But let's not get too excited; we still have a lot to cover. In future chapters you will see each of these concepts covered in greater detail. In chapter 3, you will learn how ArcCAD stores information and the relationships it creates.

Chapter

The ArcCAD Data Model

Themes, Coverages, dBASE Files, and Topology

In this chapter you will tour the ArcCAD data model. A *data model* is the method a program uses to store information. Within the ArcCAD model there are three main components: themes, coverages, and dBASE files. Through the use of coverages, ArcCAD is able to define a set of relationships between spatial features called *topology*. Direct access to dBASE files allows ArcCAD to share nongraphic-based information with many other applications.

Themes: Linking GIS Data Sets to AutoCAD

A *theme* is the definition of a GIS data set stored in an AutoCAD drawing. The theme contains the name of a data set, its location, the access method (line, point, polygon, image, record, tic, or annotation), and the symbol with which to display the spatial data. In the previous chapter you saw that themes do not actually contain any data; they simply tell AutoCAD what name you will use to access the data, what type of data to access, and where the data is located on your system. The one additional property defined in a theme is a default symbol used when displaying the coverage data in the drawing.

Themes are drawing-based entities and are stored when the AutoCAD drawing file is saved. Each time you open that drawing, ArcCAD will automatically be connected to those themes. Multiple themes may be defined within an AutoCAD drawing, and multiple AutoCAD drawings may have themes defined for the same data set. With this functionality, AutoCAD becomes an input and display device for ArcCAD coverages. Once a map has been created by displaying data from the coverages, it can be saved as an AutoCAD drawing.

Themes can be thought of as parallel to AutoCAD layers. Each logical set of data defining a project is defined as a separate theme. The following diagram shows an example of a drawing split into multiple themes.

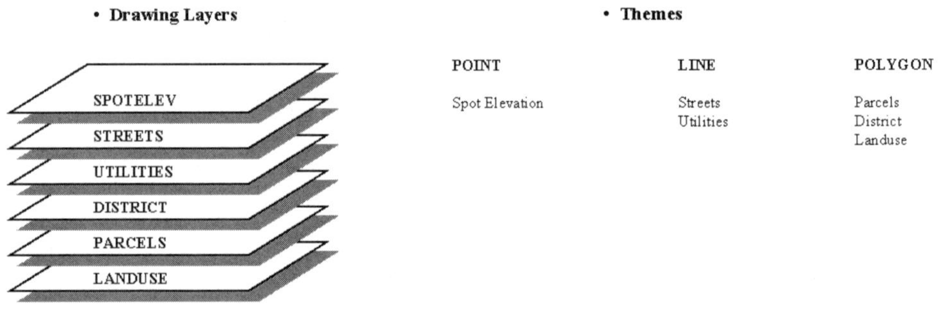

A diagram of a data set broken down into themes.

Theme Types

When creating a theme, you must specify what type of data you want to access. The seven data types are point, line, polygon, record, image, annotation, and tic. A data set must fit into at least one of these categories and some fit into more than one. For example, all polygon themes can also be accessed as line themes since polygon boundaries are, in fact, lines. The purpose of defining a type within theme creation is to identify to ArcCAD what entities to link data to. For example, in the polygon data mentioned, if you wish to attach attribute information to the area contained within the polygon boundary, you would define a polygon theme. To attach attributes to the polygon boundary itself, a line theme would be defined.

In Chapter 2, "A Sample Work Session," the soils theme was defined as polygons. This allowed you to assign the SOILS_NAME attribute to the area contained within each polygon. Let's work through an example of how you can assign attributes to the boundaries themselves.

Setup

To begin, start ArcCAD and begin a new drawing named INARC2. Be sure this drawing file is created in the INARC directory. If ArcCAD is not loaded automatically, load it now.

In order to access the soils data created in Chapter 2, you must define a theme for the data set. Execute the Defthm command found on the Themes menu. Enter SOILS as the theme name, Polygon as the theme type, and C:\INARC\SOILS as the data set name. Once the dialog box is complete, select OK. The completed dialog is shown in the following figure.

The Define Theme dialog box for the Soils theme.

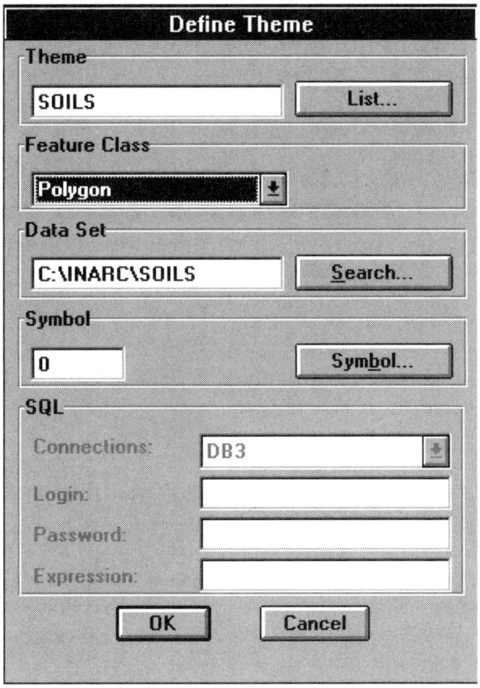

With the theme defined, you are ready to display the polygons into the drawing file. Since this is a new AutoCAD drawing, the limits have not yet been set.

To set the AutoCAD limits to the limits of the SOILS data set, select the Setext (set extents) command found on the Display menu. At the prompt for the theme name, enter SOILS and press RETURN twice. The AutoCAD drawing is zoomed out to display the new area. To display the soils into the drawing select Polys from the Display, POLYGONS menu, press Enter to accept SOILS, and press Enter to accept symbol number 0.

The polygons should now be displayed in the drawing. To experiment with this coverage, select Identify from the Query menu. Press RETURN at the theme name prompt to accept SOILS. Notice that ArcCAD prompts you to select a polygon, but does not provide a pick box for selecting an entity. It expects you to pick a point inside a polygon.

Now let's look at this same coverage defined as a line theme. To do this, define a new theme using the Defthm command found on the Themes menu. Enter SOILS_LN as the theme name, LINE as the type, and C:\INARC\SOILS as the data set name.

With the theme defined, execute the Build command found on the Topology menu to update the data set with line information. Specify SOILS_LN as both the input theme and the output theme. To view the information created for the soils boundaries, select the Wbrowse command found on the Attributes DATAFILES menu. The feature attribute table (FAT) should look like the following figure.

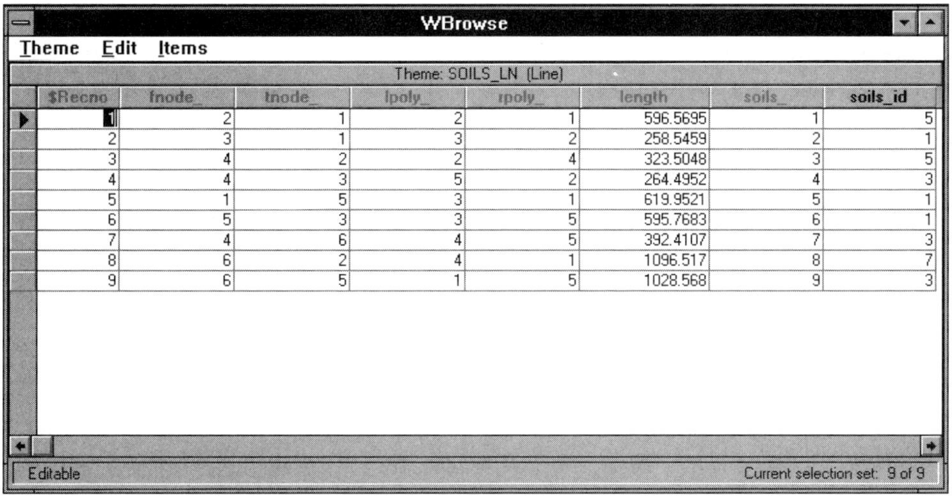

$Recno	fnode_	tnode_	lpoly_	rpoly_	length	soils_	soils_id
1	2	1	2	1	596.5695	1	5
2	3	1	3	2	258.5459	2	1
3	4	2	2	4	323.5048	3	5
4	4	3	5	2	264.4952	4	3
5	1	5	3	1	619.9521	5	1
6	5	3	3	5	595.7683	6	1
7	4	6	4	5	392.4107	7	3
8	6	2	4	1	1096.517	8	7
9	6	5	1	5	1028.568	9	3

The SOILS arc attribute table.

Notice that this FAT looks considerably different than the polygon attribute table created in the previous chapter. This table contains information for lines rather than polygon areas.

To return to the drawing and look at a specific feature, select Exit from the dialog box's Theme menu. To display the lines, you need to set the text size for labels, clear the screen, and display the lines. Follow the input below:

```
Command: e
ERASE
Select objects:200,100
Other corner: 1600,1000
9 found
Select objects:ENTER
```

Set the label text size by entering Textsize at the command prompt.

```
Command: textsize
New value for TEXTSIZE <0.2000>: 20
```

Select Arcs from the Display, LINES menu.

```
Command: arcs
Theme name (?/<SOILS_LN>): ENTER
Display IDs? (Only/Yes/<No>): yes
```

With the polygon boundaries displayed as lines in the drawing, you can now use the Identify command found on the Query menu to select a specific feature.

At the theme name prompt press Enter to accept the default of SOILS_LN. Notice now that ArcCAD prompts you to select an entity and provides a pick box. Select one of the lines. The Record Information dialog box now displays all attributes for the selected feature.

The relationship between theme types is an important one and will be discussed further in later chapters. The concept of how ArcCAD stores different coverages and the relationship between lines and polygons is discussed further in the coverages section.

Coverages

A *coverage* is one of ArcCAD's internal data storage formats. Each coverage created is stored in a separate directory on disk. These coverages are then accessed by defining a theme in an AutoCAD drawing.

A coverage contains spatial databases defining the location of the graphic portion of the database as well as at least one FAT. The FAT contains user-accessible attribute information describing the spatial features, with an internal ID that links one record in the FAT to one feature in the spatial database.

As you have seen in the previous section, a coverage may contain more than one FAT. In the previous example, the soils coverage contained both a polygon attribute table and an arc (line) attribute table. The polygon attribute table contained one record of attribute information for each of the four polygons. The arc attribute table contained one record for each of the nine lines bounding the polygons.

To look at the SOILS data as it is stored, type `SHELL` at the command prompt and press Enter twice.

At the DOS prompt, change directory to the INARC directory if necessary and list the SOILS subdirectory.

```
C:\INARC>DIR SOILS
```

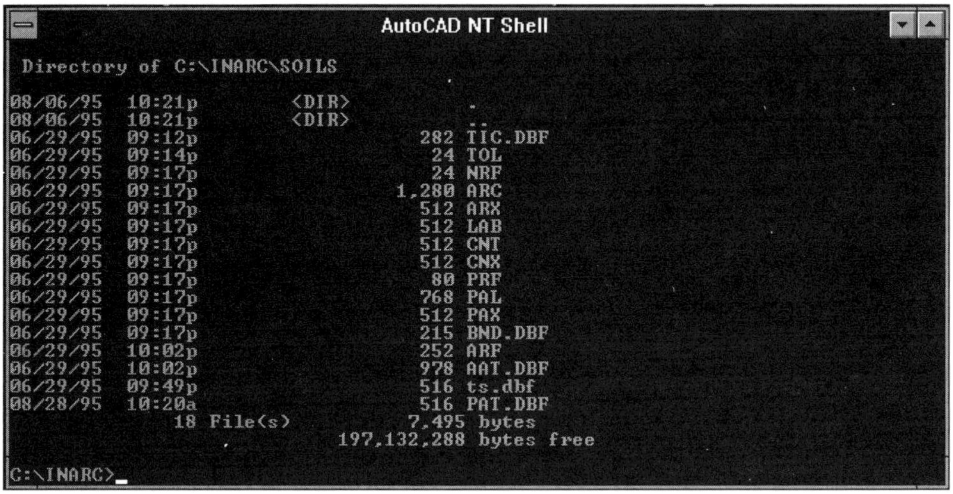

Directory listing of the SOILS coverage.

As you can see, the directory contains both a polygon attribute table (PAT.DBF) and an arc attribute table (AAT.DBF), as well as a series of spatial databases.

To return to the drawing editor type EXIT at the DOS prompt.

Within a coverage, a point feature will be stored as a single x,y coordinate pair. Line features are stored as a series of x,y coordinate pairs and polygon features (homogeneous areas enclosed by the lines that bound them) are stored as a series of arcs that form their boundaries.

The PAT and AAT feature attribute tables are stored in industry standard DBF format and can be accessed with any application capable of reading DBF format. The spatial database files are in a binary format and are not accessible to external applications.

dBASE Files

dBASE III formatted files are used to hold attribute information in ArcCAD. They differ from coverages in that they contain only attribute information, whereas coverages hold both attribute data as well as coordinate or spatial data. A Record theme is used to access dBASE files from within ArcCAD. Record themes work identically to other theme types. A theme name is entered with a type of RECORD and the data set name is the path to and name of the dBASE file. In Chapter 2 you saw several examples using dBASE files. The first example was the SOILS lookup table. The second was the DRAINLVL datafile. It is typical for a dBASE file to hold attribute information that relates to spatial data held in other coverages.

The power of using dBASE files to hold attribute information can be seen in a parcel mapping system. The tax information for each parcel may already exist in a database or spreadsheet file. This data can be converted to dBASE format and easily related to the parcel coverage. This simplifies the process of adding existing information to your ArcCAD system. dBASE files are also beneficial in situations in which multiple users wish to access the parcel information for different purposes. Since the tax information is only temporarily related to the FAT, other users of the system are not forced to display the information when it is not needed.

Most PC-based database and spreadsheet applications have the ability to output dBASE format files. If the application you are working with can't directly write dBASE files, try exporting the information in comma delimited format. ArcCAD provides a utility called *LoadDB* for reading these comma delimited files. For more information on LoadDB, refer to the *ArcCAD Command Reference.*

Some other uses for dBASE files are as lookup tables and property tables. You have seen one example of a lookup table in Chapter 2. We will discuss lookup tables and property tables in more depth in later chapters.

Topology

So far you have learned that a spatial GIS data set is made up of information defined as points, lines, or polygons. These spatial features have a database or attribute table describing them, and the two tables are linked by a common unique identifier known as an internal ID.

Topology is a mathematical procedure for defining spatial relationships between features in a GIS data set. In a topological data model, lines connect to each other at nodes (connectivity), areas are defined by the lines that make up their boundaries (area), and lines have direction, *from* nodes and *to* nodes, as well as left and right sides (contiguity). Using topology a GIS data set can tell you which polygons are adjacent to a select polygon or which lines connect to the selected line.

If we look at the arc attribute table shown earlier, you can see that the table shows the *from* node and *to* node (direction of each line) as well as the left and right polygon for each line.

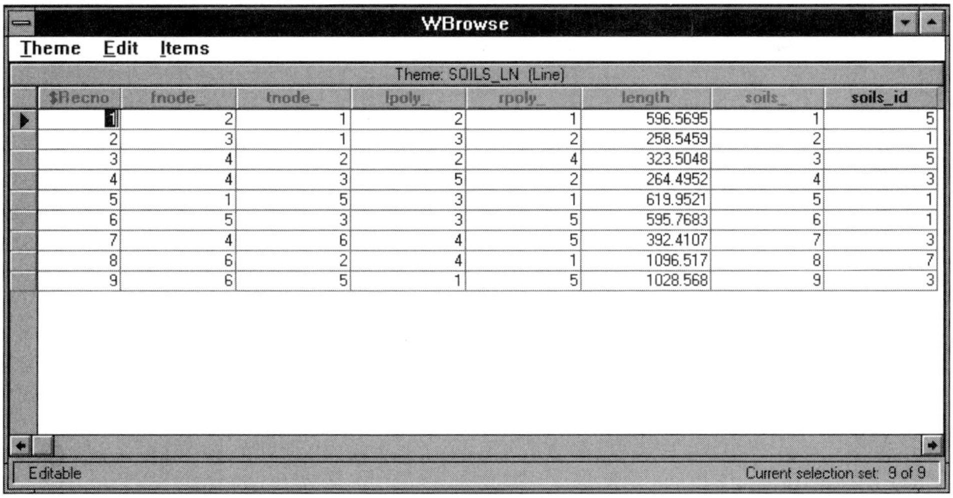
WBrowse

Theme Edit Items

Theme: SOILS_LN (Line)

$Recno	fnode_	tnode_	lpoly_	rpoly_	length	soils_	soils_id
1	2	1	2	1	596.5695	1	5
2	3	1	3	2	258.5459	2	1
3	4	2	2	4	323.5048	3	5
4	4	3	5	2	264.4952	4	3
5	1	5	3	1	619.9521	5	1
6	5	3	3	5	595.7683	6	1
7	4	6	4	5	392.4107	7	3
8	6	2	4	1	1096.517	8	7
9	6	5	1	5	1028.568	9	3

Editable

Current selection set: 9 of 9

The SOILS arc attribute table.

If you envision that the spatial databases for polygon data contain a list of the lines that make up each polygon's boundary, you can see the relationships of direction, connectivity, area, perimeter, and adjacency/contiguity.

Two Types of Topology

ArcCAD creates different types of topology, depending on the feature class of the theme. The two types of topology are arc-node topology and polygon-arc topology. When building topology for a line theme, ArcCAD will create arc-node topology, that is, ArcCAD will create an arc attribute table containing a *from* node and a *to* node, defining direction, length, and connectivity.

The *from* node and *to* node for the coverage are calculated based on how the information was digitized. For example, when digitizing a line, the first point picked will be the *from* node and the second point picked will be the *to* node. The order in which lines are digitized is only important if you need directional modeling, such as flow in pipes or rivers. The following table shows arc-node topology.

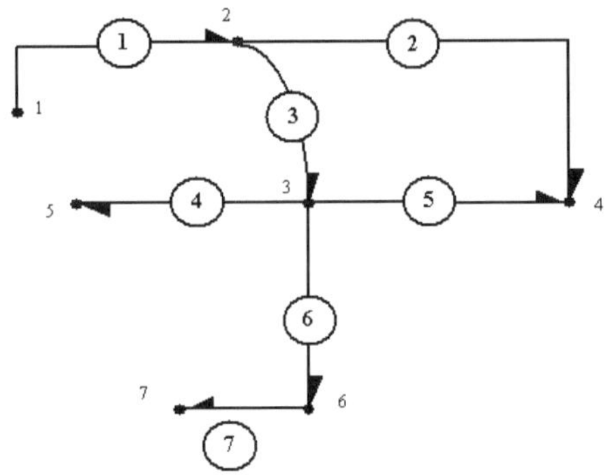

- From-node and to-node
- Direction
- Connectivity

Arc Coordinate Data

ARC#	X,Y PAIRS
1	
2	
3	
4	
5	
6	
7	x,y x,y x,y

ARC #	FNODE_	TNODE_	LENGTH
1	1	2	100.35
2	2	4	101.23
3	2	3	58.97
4	3	5	62.76
5	3	4	49.08
6	3	6	45.87
7	6	7	33.67

Arc-node Topology

A diagram of arc-node topology.

When creating topology for a polygon coverage, ArcCAD builds a polygon attribute table containing polygon-arc topology. Polygon-arc topology defines the relationship between polygon label points and the lines that bound it. This information defines area, perimeter length, adjacency, and a new concept called the *universe polygon.*

The length of the lines bounding the polygon equals the perimeter length. The area within the bounding lines defines the area of the polygon. Two polygons sharing a single arc define adjacency. The universe polygon is a polygon that is made up of all arcs defining the outer perimeter of the polygon themes features. This polygon always has a negative area that is the sum of the areas of all other polygons in the theme, and it always has an internal ID number of 1. The following table shows polygon-arc topology.

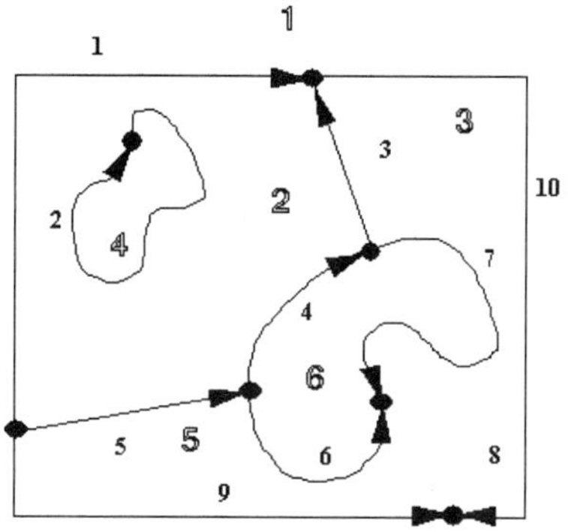

- Area Definition
- Adjacency
- Universe Polygon

Arc Coordinate Data

ARC #	X,Y PAIRS
1	
2	
3	
4	
5	
6	
7	
8	
9	
10	

ARC #	LPOLY	RPOLY
1	1	2
2	2	4
3	2	3
4	2	6
5	2	5
6	6	5
7	3	6
8	3	5
9	5	1
10	1	3

Left/Right List

POLY #	ARC #
1	1,10,9
2	1,3,4,5,0,2
3	3,10,8,7
4	2
5	5,6,8,9
6	4,7,6

Polygon-arc Topology

A diagram of polygon-arc topology.

How ArcCAD Creates Topology

Topology is created for a coverage by either cleaning or building the data set. Through the Clean or Build command, a FAT is created with the items necessary to define topology.

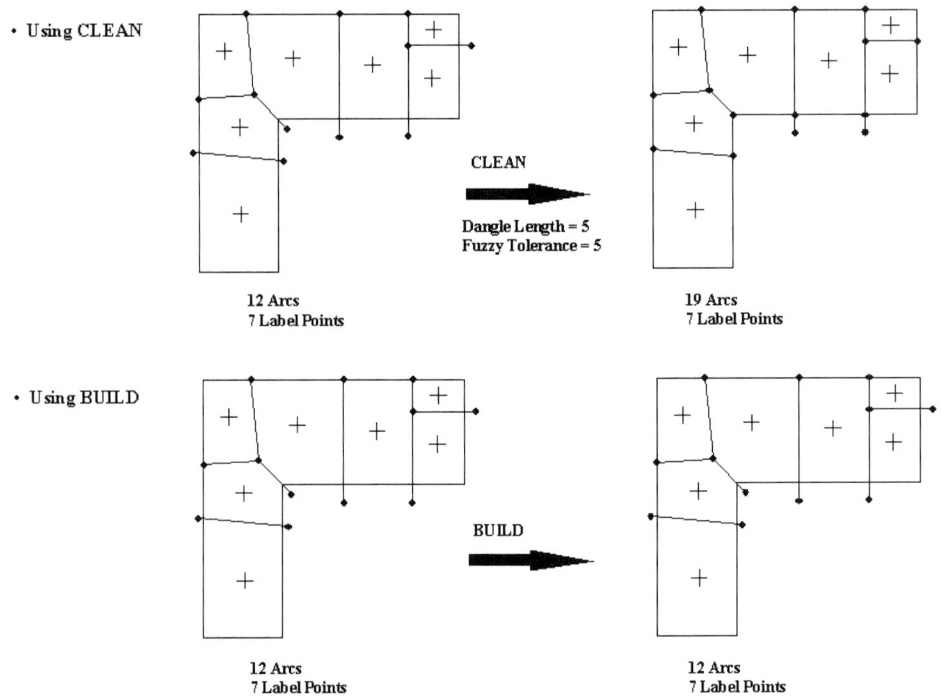

A diagram of the Clean and Build commands.

The Build and Clean commands are similar in that they both create topology. The Clean command will detect and split arcs that intersect, adjust node (arc endpoints) so that two arcs with endpoints within a specified tolerance (fuzzy) will connect, and delete lines that are less than a specified length (dangle). Build, on the other hand, will not split arc intersections, connect arc endpoints, or delete unwanted lines.

The other major difference between Build and Clean is that Clean works only on line and polygon data types, whereas Build works on point, line, and polygon data types. The following table shows the uses of Build and Clean. The following table shows how point, line, and polygon themes will be affected by Clean and Build.

Characteristics	Build	Clean
Constructs topology		
Polygons and a PAT	✓	✓
Lines and a AAT	✓	if exist
Points and a PAT	✓	
Establishes or resets internal numbers		
Polygons	✓	3
Lines	✓	3
Calculates spatial measurements	✓	3
Creates intersections		✓
Works if arcs cross		
Polygons		✓
Lines	✓	3
Corrects errors		
Collapses colinear arcs		✓
Collapses neighboring nodes		✓
Removes dangling arcs		✓
Speed of processing	Faster	Slower

Feature Attribute Tables

Feature attribute tables (FATs), the PAT.DBF and AAT.DBF files mentioned previously, are dBASE III files. They hold ArcCAD-defined and user-defined database information that define topology and describe the spatial features in a GIS data set. These files are created whenever topology is built for a coverage. The format of the file is column headings called Items and rows of information called Records.

For each feature in the data set there is one record in the attribute table. The table record is related to the spatial feature by a unique identifier called the internal ID. This ID field is created by ArcCAD and is used only by ArcCAD. You should never modify the Internal_ID. Doing so will corrupt the link between the attribute table record and the spatial feature it represents.

The internal ID's column or item name is derived by appending an underscore (_) to the coverage name. For example, in the soils coverage the internal ID is SOILS_. The second item that is constant in all FATs is the User-ID. This item is a unique number assigned to each record in the table. The name of the item is derived by appending _ID (underscore ID) to the coverage name or SOILS_ID in the soils coverage. The User-ID allows you to identify which record is associated with each feature in the data set. You saw an example of this in Chapter 2.

The following graphics show the four types of FATs created by ArcCAD, with their standard items.

Polygon	Area	Perimeter	Cov_ID
Point	Area	Perimeter	Cov_ID
Line	Fnode_	Tnode_	Rpoly_ Length Cov_ Cov_ID

New items can be added to FATs using either ArcCAD or any external application that can read DBF format files. When adding user-defined items to an attribute table, the position and order of the standard items should never be changed, that is, no user-defined item should ever be placed to the left of the User-ID. Doing so could corrupt the relationship between the attribute table and the features it represents. Although an external application such as dBASE or Excel is not needed to run ArcCAD, it is recommended to speed the input of large amounts of attribute data.

Sources of Data

This section introduces you to the types of data available to ArcCAD and more specifically the types of data you will find in the sample projects to be completed throughout this book. This section also defines the relationship between AutoCAD entities and ArcCAD features.

ArcCAD-compatible GIS data sets include ARC/INFO and PC ARC/INFO coverages, dBASE, or dBASE-compatible databases, AutoCAD or compatible drawing files, several raster image formats, and most SQL databases.

AutoCAD Drawings

AutoCAD drawing files contain information that can easily be written to ArcCAD coverages. The AutoCAD entities readable by ArcCAD are dependent on what type of theme is being created. The following table shows what AutoCAD entities may be added to each type of ArcCAD theme.

- **Point features**

  Attributes
  Block (insert)
  Point
  Shape
  Text

- **Line features**

  Arc
  Circle
  Line
  Polygon
  Polyline
  Solid
  Trace
  3DFace

- **Polygon features**

Boundary	Label point
Arc	Attributes
Circle	Block (insert)
Line	Point
Polygon	Shape
Polyline	Text
Solid	
Trace	
3DFace	

- **Annotation features**

  Text

Relationship between AutoCAD entities and ArcCAD features.

DXF Files

Drawing Interchange Files (DXF) are ASCII files written and read by most computer-aided design and drafting systems. DXF files are typically used to transfer drawing files from one CAD system to another. Through AutoCAD's ability to read DXF files, essentially all CAD data is available for use in ArcCAD. To bring a DXF file into AutoCAD, execute the `DXFIN` command at the command prompt. The sample project used in this book contains a DXF file of the land use information. In Chapter 4 this information will be imported.

PC ARC/INFO Coverages

PC ARC/INFO coverages are identical to ArcCAD coverages. To use a PC ARC/INFO coverage in ArcCAD, define a theme identifying the coverage to AutoCAD. When using a PC ARC/INFO coverage, all spatial and attribute data is accessible, and topology is typically already defined. However, it is possible for a coverage to exist without topology. For example, the information may have been digitized in PC ARC/INFO but not cleaned or built. If topology does not exist, use the Clean or Build command in ArcCAD on the coverage.

The parcels information in the project data set is a PC ARC/INFO coverage. In the next chapter you will analyze this data set.

ARC/INFO Export Files

Although ARC/INFO, ESRI's workstation-based GIS system, has the ability to create DXF files, there is a better way to transfer data from ARC/INFO to ArcCAD. ARC/INFO can transfer its coverages to exchange files readable by ArcCAD. These files have the format of *filename.E00* through *filename.Exx* where *xx* is the total number of files created. For example, to write a specified coverage to export files with 40,000 lines per file may take four files. These files would be named *filename.E00* to *filename.E04.* Export files are better than DXF files because they allow the transfer of spatial information, graphics, and the feature attributes or database.

Once the export files have been created from ARC/INFO, the ArcCAD Import command is used to convert them to ArcCAD coverages. Themes are then defined to allow AutoCAD access to these coverages, and the data may be queried, analyzed, and displayed.

ArcCAD not only has the ability to read ARC/INFO export files, it can also create export files for transport back to ARC/INFO. This allows for a great amount of flexibility between users of ARC/INFO and ArcCAD.

The project's building information is contained in a single export filename *BUILDING.E00.* In the following chapter you will import this file and use it in your analysis.

Review

The following are the basic concepts you have learned in this chapter:

- ❒ The ArcCAD menu structure.
- ❒ How ArcCAD uses dialog boxes.
- ❒ GIS data sets.
- ❒ How ArcCAD stores GIS data sets as coverages.
- ❒ How AutoCAD gains access to this data using themes.
- ❒ Topology.
- ❒ Sources of Data.

In later chapters you will learn how themes, coverages, and topology are created, and you will use this information to complete a GIS project.

Chapter

Defining the GIS Project

Questioning, Analyzing, and Mapping

You have seen how ArcCAD stores data, how these data sets are associated with an AutoCAD drawing, and how ArcCAD can analyze and display GIS data sets. In this chapter we discuss the major steps in a GIS project and define the sample project to be completed throughout this book.

The first step in defining the GIS project is to identify the questions that must be answered. With this information, you can begin to define what data layers must be created to achieve the desired results. Once the data has been compiled, the analysis can begin. Finally, the map displays and reports can be generated. There are many excellent books on defining a GIS if you wish to explore this topic in more detail.

Building Geographic Data

In this chapter you will learn to assemble spatial data, evaluate the data for errors, correct the errors, add attribute data to the spatial information, and use relate tables to define a complete GIS data set.

All of the data created in Part I will be used in subsequent chapters. If you have trouble creating any of the data sets, they can be copied from the C:\INARC\COMPLETE\> directory.

Assembling Spatial Data

Assembling spatial data entails identifying and organizing the logical sets of data necessary for a GIS project. This process includes analyzing the data formats available and the method of converting these formats to ArcCAD coverages.

In this section you will define six coverages and a series of lookup and relate tables.

The coverages and their sources are defined below.

- Utility poles and lines, soils and trees: AutoCAD drawing files
- Landuse polygons: DXF file
- Buildings: ARC/ INFO export files
- Streets and parcels: PC ARC/INFO coverages

In building the GIS coverages, the following commands will be used:

- **Defthm:** Define theme
- **Ddtheme:** Themes menu dialog box
- **UserID:** Set the current User-ID base value
- **Addfeat:** Add features to a theme
- **Build:** Build topology for a theme
- **Clean:** Automatically correct spatial errors and build topology for a theme
- **Createlab:** Create polygon label points for a polygon theme
- **Laberror:** Identify polygon label point errors
- **Noderror:** Identify arc node errors

Creating a Coverage from AutoCAD Entities

You will begin the process of building geographic data sets by using the ArcCAD tools for converting AutoCAD drawing entities to ArcCAD coverages. These tools allow you to convert graphic entities to spatial data, and allow you to extract AutoCAD properties and write them to the data set's feature attribute table (FAT).

To convert AutoCAD entities into an ArcCAD coverage, a theme is defined identifying a new coverage. The Addfeat command is used to write the entities to the coverage and the Clean and Build commands are used to create topology for the coverage. During the process of writing AutoCAD entities to a theme, an optional table can be used to extract AutoCAD properties. The AutoCAD properties that may be extracted are listed below.

Color The AutoCAD color number for this entity. Color is written to an item named *acad_color*.

Layer name The layer name that the entity is on. Layer name is written to an item named *acad_layer*.

Linetype The linetype of the entity. Linetype is written to an item named *acad_ltype*.

Elevation The elevation of the entity. Elevation is written to an item named *acad_elev*.

Thickness The thickness of the entity. Thickness is written to an item named *acad_thick*.

Entity type A character string describing the entity type. Entity type is written to an item named *acad_etype*.

Entity handle A character string containing the entity handle. Entity handle is written to an item named *acad_hand*.

Curve information A number indicating whether the entity is a true curve. A "1" indicates a true curve, a "0" indicates this entity does not have curve information. Curve information is written to an item named *acad_curve.*

Text The text of a text entity. Text is written to an item named *acad_text.*

Block The block name of the insert.

Attributes Attribute information stored in a block. The ITEM name is the attribute's tag name.

You will use optional property tables to automate the process of building a GIS data set.

Line Themes

The first set of AutoCAD entities you will write to an ArcCAD coverage are the project's Utility Lines.

Setup

To begin, start ArcCAD and open the INARC3 drawing. If ArcCAD has not been configured to start automatically, start it now by entering "Load arcad" at the AutoCAD command prompt.

The INARC3 drawing file should look like the following figure.

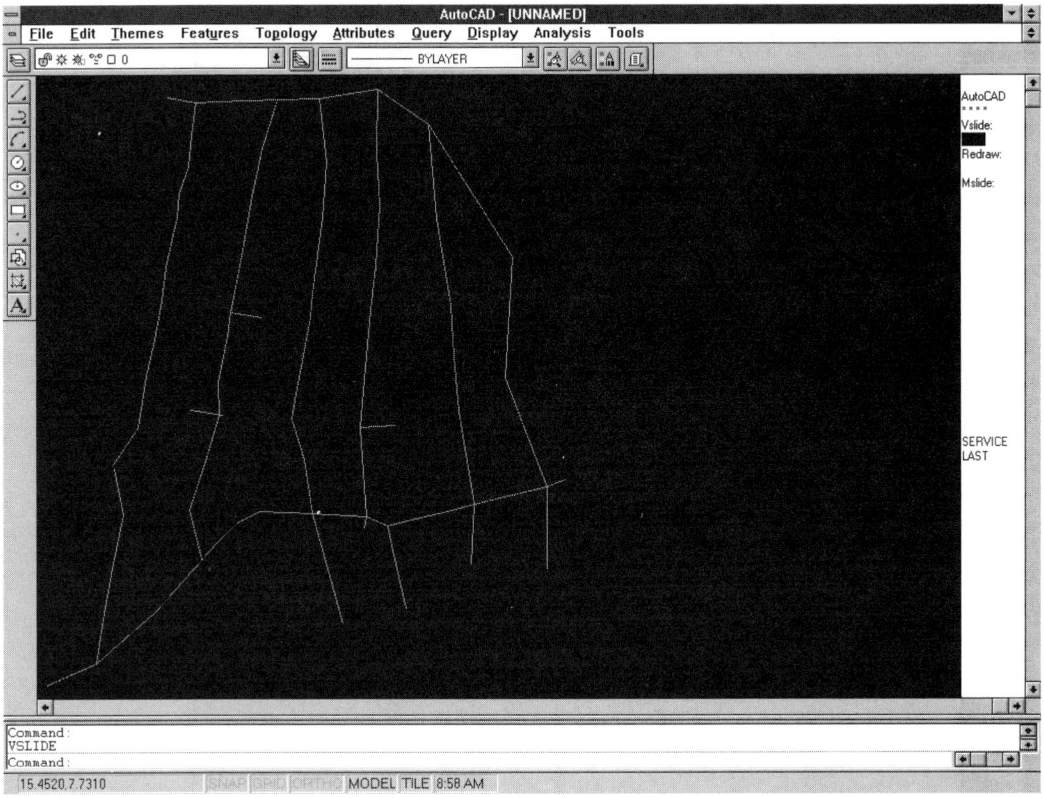

The INARC3 drawing file.

Before moving on to defining the UTILITY LINES theme, use the AutoCAD List command to list several of the lines in the drawing. You should notice that there is a mix of lines and polylines making up the drawing. For example, the entire right utility line is made up of lines, whereas the utility lines running down the middle are polylines. Once you have completed looking through the drawing, define the UTILITY LINES theme by selecting Defthm from the Themes menu and filling out the dialog box as shown below.

```
Command: DEFTHM
```

The Define Theme dialog box for the Utility lines theme.

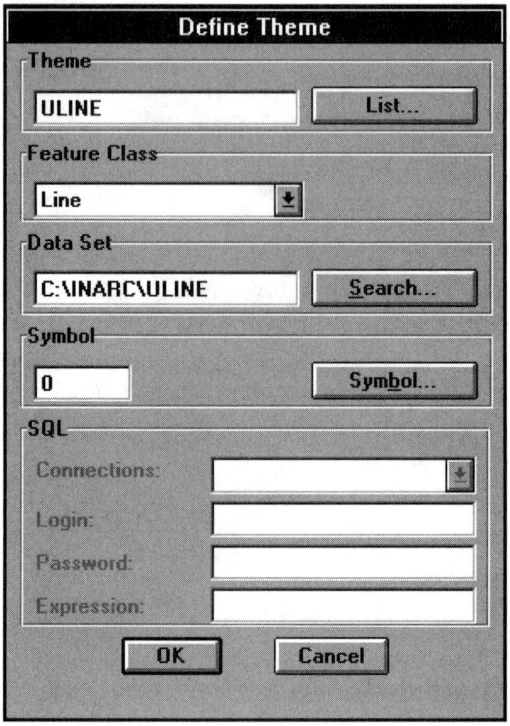

Now use the Features menu to add features to the data set.

To add features to your Utilities theme, select Addfeat from the Features pulldown menu.

At the Input theme prompt press ENTER to accept the default theme of Uline.

Addfeat now requests the user to select objects. Any AutoCAD object selection method may be used here, including filters. Use a window to select all the utility lines, as shown in the following figure.

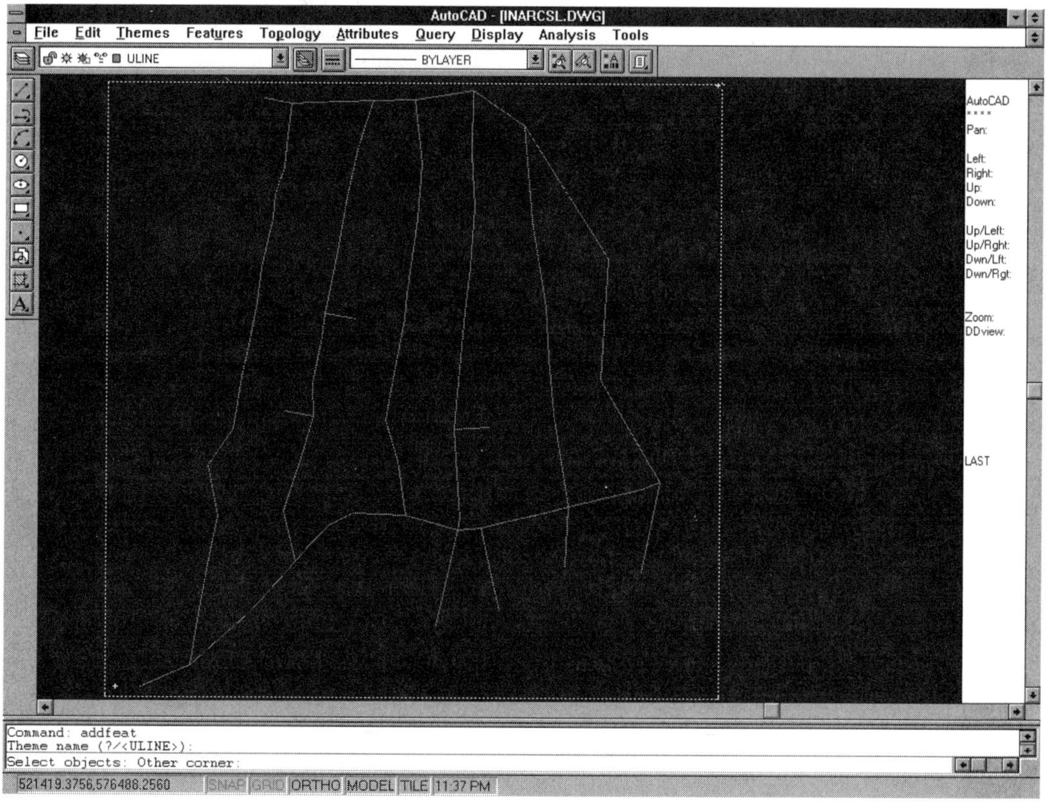

This figure shows a snapshot with the window drawn dotted.

Once all entities are selected, press ENTER at the Select Objects prompt. ArcCAD has now written all of the features to the coverage, Addfeat will display a dialog box, asking whether to Build, Clean, or Wait.

Build Versus Clean

Build and Clean are the two ArcCAD commands that create topology. The commands are similar, but have a few major differences. The most significant difference is that Clean splits all intersecting arcs, deletes dangling arcs (arcs that extend into space less than a user-specified dangle length), and snaps nodes that fall within a user-specified tolerance together to form closed polygons and line intersections. The Build command makes no modifications to the spatial data. The Build command processes point, line, and polygon themes, and the Clean command processes only line and polygon themes.

You will take a closer look at these differences by using both commands on the ULINES theme and comparing the results.

Select Build from the dialog box to build the ULINE theme now.

The Clean/Build dialog box.

Building the theme created topology for the utility lines without splitting intersections or snapping nodes together. If we look at the attribute table for the ULINE theme, you will notice that ArcCAD also did not add *from* node or *to* node, or left and right polygon information to the table.

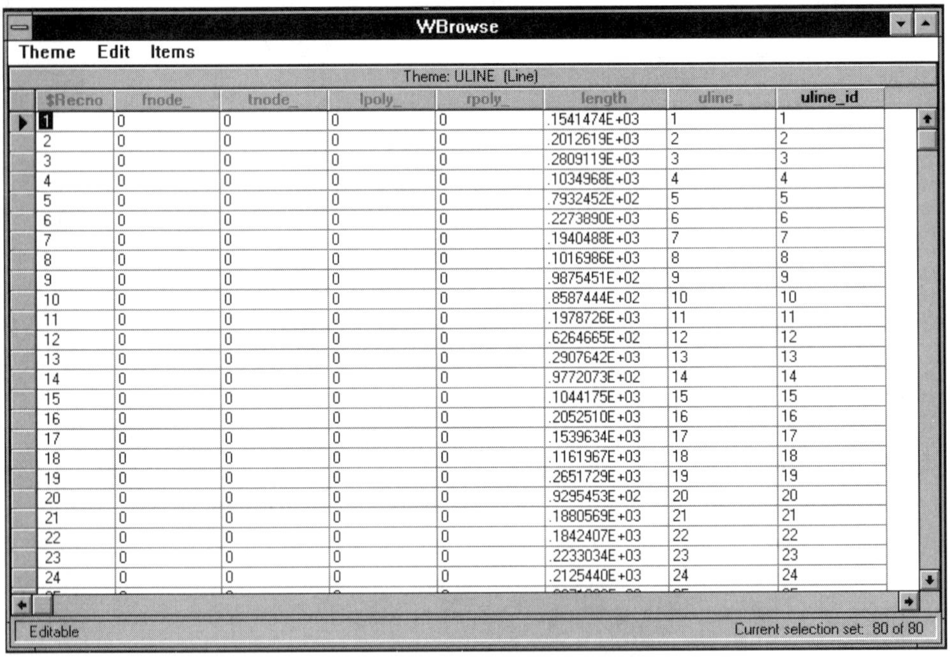

$Recno	fnode_	tnode_	lpoly_	rpoly_	length	uline_	uline_id
1	0	0	0	0	.1541474E+03	1	1
2	0	0	0	0	.2012619E+03	2	2
3	0	0	0	0	.2809119E+03	3	3
4	0	0	0	0	.1034968E+03	4	4
5	0	0	0	0	.7932452E+02	5	5
6	0	0	0	0	.2273890E+03	6	6
7	0	0	0	0	.1940488E+03	7	7
8	0	0	0	0	.1016986E+03	8	8
9	0	0	0	0	.9875451E+02	9	9
10	0	0	0	0	.8587444E+02	10	10
11	0	0	0	0	.1978726E+03	11	11
12	0	0	0	0	.6264665E+02	12	12
13	0	0	0	0	.2907642E+03	13	13
14	0	0	0	0	.9772073E+02	14	14
15	0	0	0	0	.1044175E+03	15	15
16	0	0	0	0	.2052510E+03	16	16
17	0	0	0	0	.1539634E+03	17	17
18	0	0	0	0	.1161967E+03	18	18
19	0	0	0	0	.2651729E+03	19	19
20	0	0	0	0	.9295453E+02	20	20
21	0	0	0	0	.1880569E+03	21	21
22	0	0	0	0	.1842407E+03	22	22
23	0	0	0	0	.2233034E+03	23	23
24	0	0	0	0	.2125440E+03	24	24

The ULINE FAT.

The Clean command always builds this topology information into the table. To create the *from* node and *to* node topology without cleaning (splitting intersections), use the Renode command found on the Topology menu.

```
Command: renode
Theme name (?/<ULINE>): ENTER
     76 unique nodes built.
```

The FAT now contains the *from* node and *to* node information, identifying the direction of each line.

WBrowse

Theme Edit Items

Theme: ULINE (Line)

$Recno	fnode_	tnode_	lpoly_	rpoly_	length	uline_	uline_id
1	4	3	0	0	.1541474E+03	1	1
2	62	56	0	0	.2012619E+03	2	2
3	24	16	0	0	.2809119E+03	3	3
4	2	6	0	0	.1034968E+03	4	4
5	6	5	0	0	.7932452E+02	5	5
6	5	4	0	0	.2273890E+03	6	6
7	56	55	0	0	.1940488E+03	7	7
8	55	60	0	0	.1016986E+03	8	8
9	60	63	0	0	.9875451E+02	9	9
10	63	65	0	0	.8587444E+02	10	10
11	65	69	0	0	.1978726E+03	11	11
12	69	71	0	0	.6264665E+02	12	12
13	71	74	0	0	.2907642E+03	13	13
14	74	75	0	0	.9772073E+02	14	14
15	75	76	0	0	.1044175E+03	15	15
16	56	47	0	0	.2052510E+03	16	16
17	47	43	0	0	.1539634E+03	17	17
18	43	38	0	0	.1161967E+03	18	18
19	38	30	0	0	.2651729E+03	19	19
20	30	24	0	0	.9295453E+02	20	20
21	4	9	0	0	.1880569E+03	21	21
22	9	14	0	0	.1842407E+03	22	22
23	14	20	0	0	.2233034E+03	23	23
24	20	28	0	0	.2125440E+03	24	24
25	28	31	0	0	.8971032E+02	25	25
26	31	37	0	0	.1709357E+03	26	26
27	37	42	0	0	.1226081E+03	27	27

Editable — Current selection set: 80 of 80

The Uline FAT after using the Renode command.

To examine the spatial information for the ULINE theme, display the utility line data into the drawing file onto a new layer.

```
Command: LA
```

```
?/Make/Set/New/ON/OFF/Color/Ltype/Freeze/Thaw/
  LOck/Unlock: M
New current layer <0>: WORK
?/Make/Set/New/ON/OFF/Color/Ltype/Freeze/Thaw/LOck/
  Unlock:OF
Layer name(s) to turn Off: *
Really want layer WORK (the CURRENT layer) off?  <N>N
?/Make/Set/New/ON/OFF/Color/Ltype/Freeze/Thaw/LOck/
  Unlock: ENTER

Command: ARCS
Theme name (?/<uline>): ENTER
Display IDs? (Only/Yes/<No>): YES
```

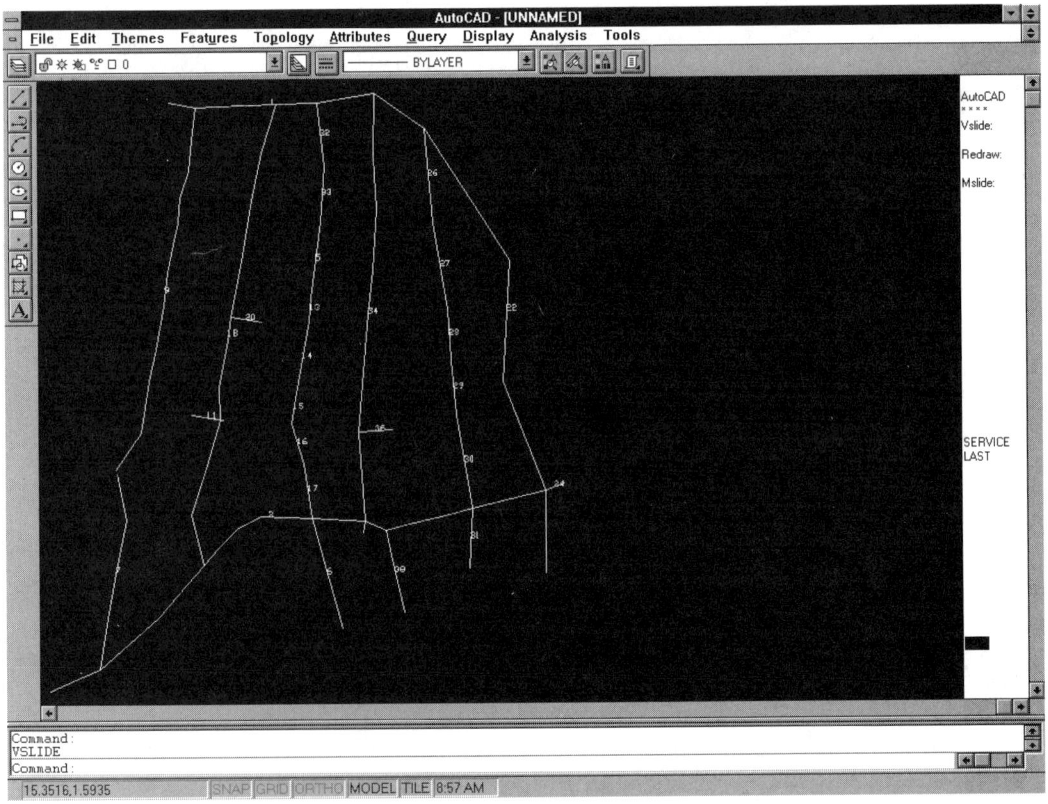

The Utility lines with User-IDs.

Careful inspection of the drawing will show that ArcCAD did not break the utility line segments at the intersections or snap endpoints together.

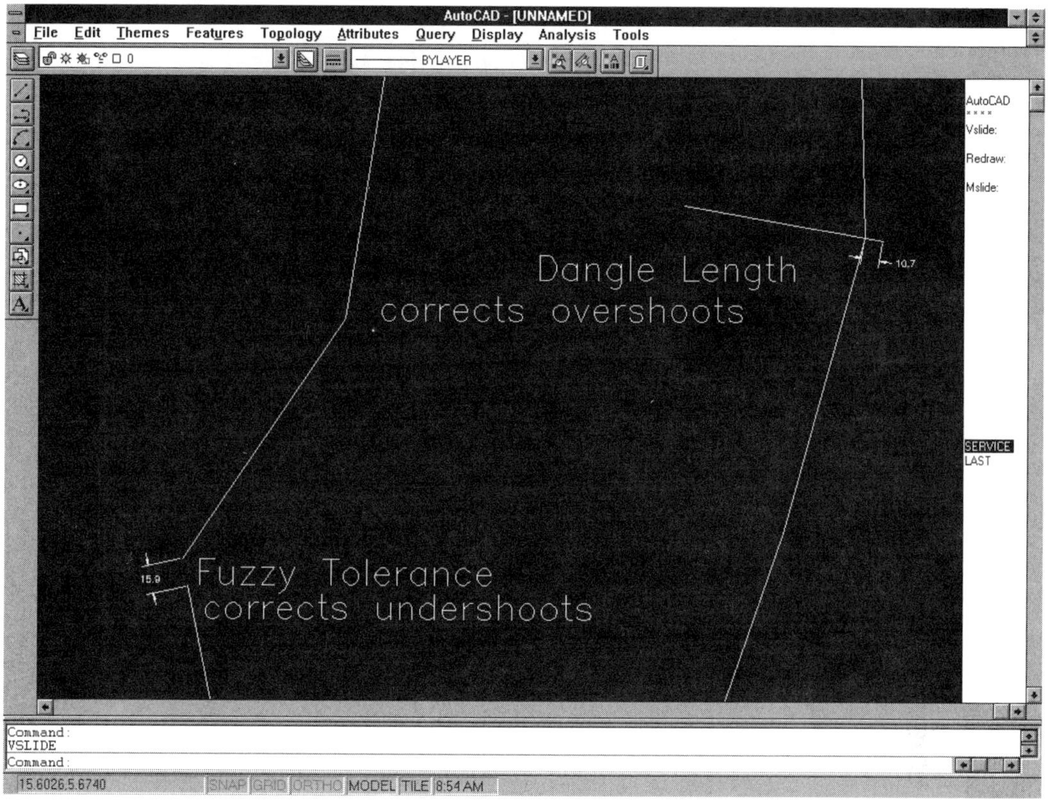

Nonjoining endpoints and dangles.

You should also note that lines drawn as AutoCAD polylines are displayed from the coverage as continuous polyline segments with nodes at their beginning and end. Lines that were drawn as AutoCAD lines were returned as polylines, with nodes at each original line endpoint.

If you only wanted certain intersections broken, you would have to draft so that lines start and end at intersections as desired. If the desired result is to have all intersections broken, the Clean command can be used.

To demonstrate, Clean the ULINE theme to a new theme. The new theme will then be built to update the topology.

Select Clean from the Topology menu. At the theme name prompt, enter Uline and follow the input below.

```
Command: clean
Input theme (?/<ULINE>):
Output theme (?/<ULINE>): ulinec

Theme ULINEC is not defined.
```

Select OK to define the new theme

```
Defining ULINEC...
Dangle length <default>:20
Fuzzy tolerance <compute>:20
Cleaning ULINE.
Copying ULINE to ULINEC...
Sorting...
Intersecting...
Assembling polygons...
  Sorting input file...
  Sorting label file...
  Processing...
  Assigning final IDs...
  Writing arc file...
  Generating polygon report...
Creating Polygon attribute table.
Re-building arc attribute table...
Creating Line attribute table.
```

Clean Fuzzy and Dangle

Clean is an ArcCAD command designed to review a set of spatial information for errors, make some determinations on how to correct the errors, and then modify the spatial data to eliminate the error. To avoid having the Clean command adversely affect data sets by making unnecessary modifications, two tolerances may be applied to the command. These two tolerances are the Fuzzy Tolerance and the Dangle Length.

Fuzzy tolerance is a value that specifies the minimum distance that can exist between two points. If two points are separated by less than this value, the two points will be collapsed together into a single location.

Dangle length specifies the minimum length all individual unconnected line segments are compared. If the line segments are less than the dangle length, the line segment will be removed during the clean process.

Care should be taken when specifying fuzzy tolerances and dangle lengths. Improper values can snap nodes and delete lines incorrectly.

The following diagram shows examples of valid and invalid clean tolerances.

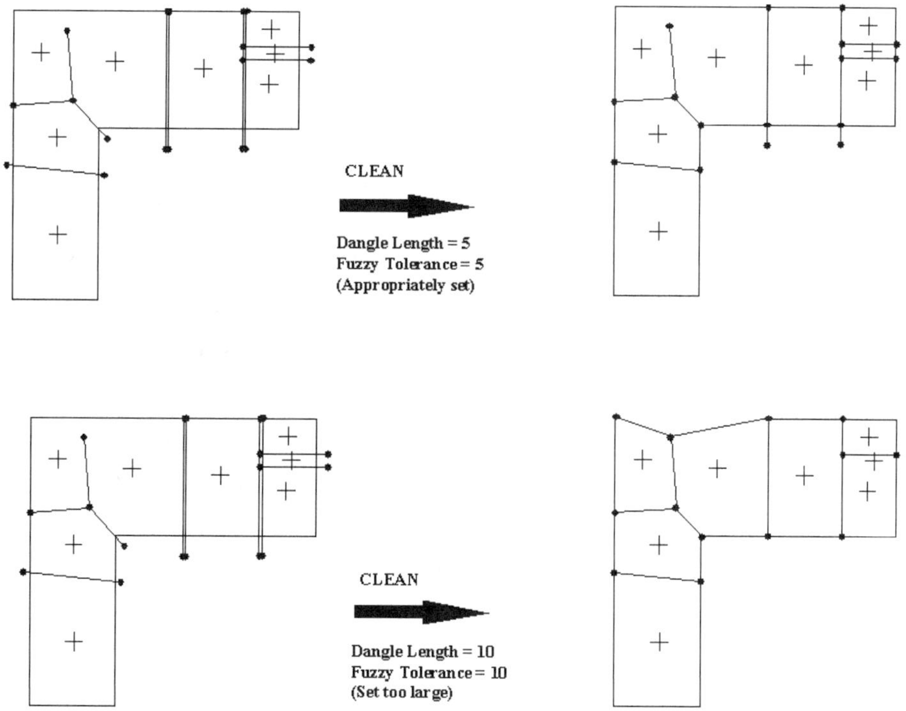

Clean tolerances.

The ULINE theme has now been cleaned to correct line endpoints that did not connect and to split intersections.

Before displaying the new theme into the drawing, erase the previous set from the drawing using the AutoCAD Erase command.

To display the cleaned utility lines in the drawing, execute the Arclines command found on the Display Lines menu.

```
Command: arclines
Theme name (?/<ulinec>):ENTER
Display using? (Properties/Item/<Symbol>):ENTER
Line symbol number (0-100) (?/<0>): 3
```

Studying the new lines will reveal that all intersections have been split, and that any nodes within a 20-foot fuzzy tolerance of each other have been snapped together.

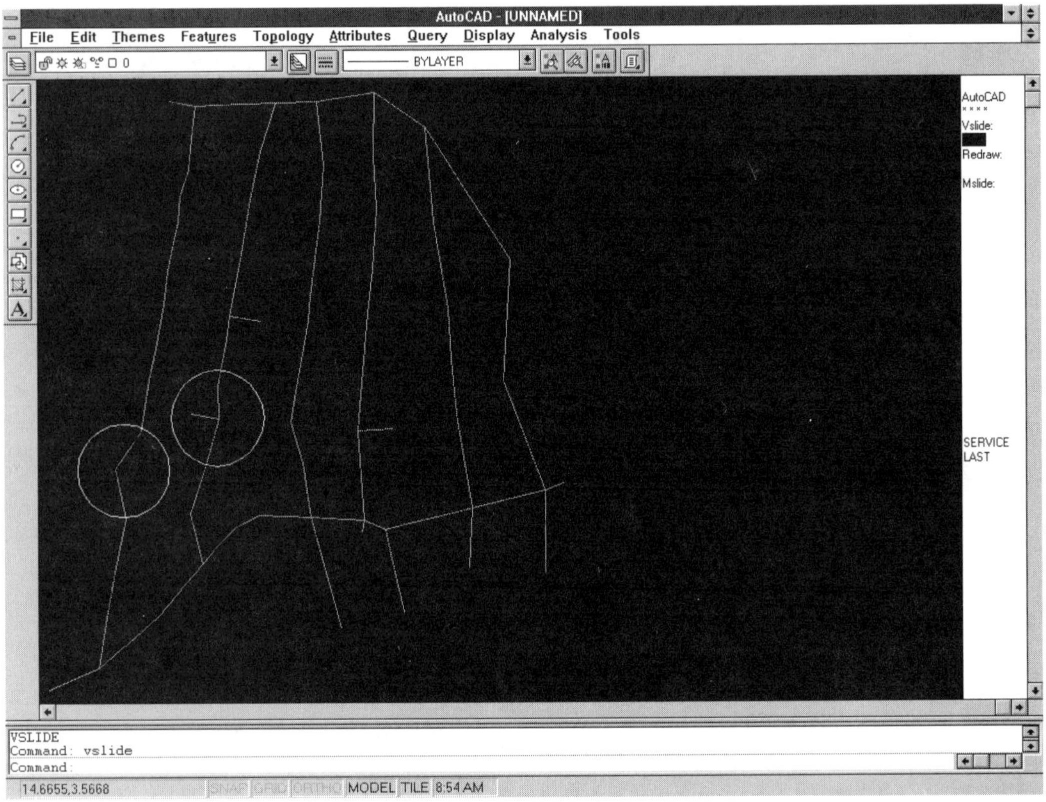

A diagram showing the results of the Clean command.

The previous command set created a spatial coverage from AutoCAD entities. This command structure wrote the AutoCAD entities to a theme called ULINE and demonstrated several ways to build topology for the data set. By listing the database using Wbrowse from the Attributes menu, you can review the attributes that ArcCAD automatically creates for a line theme.

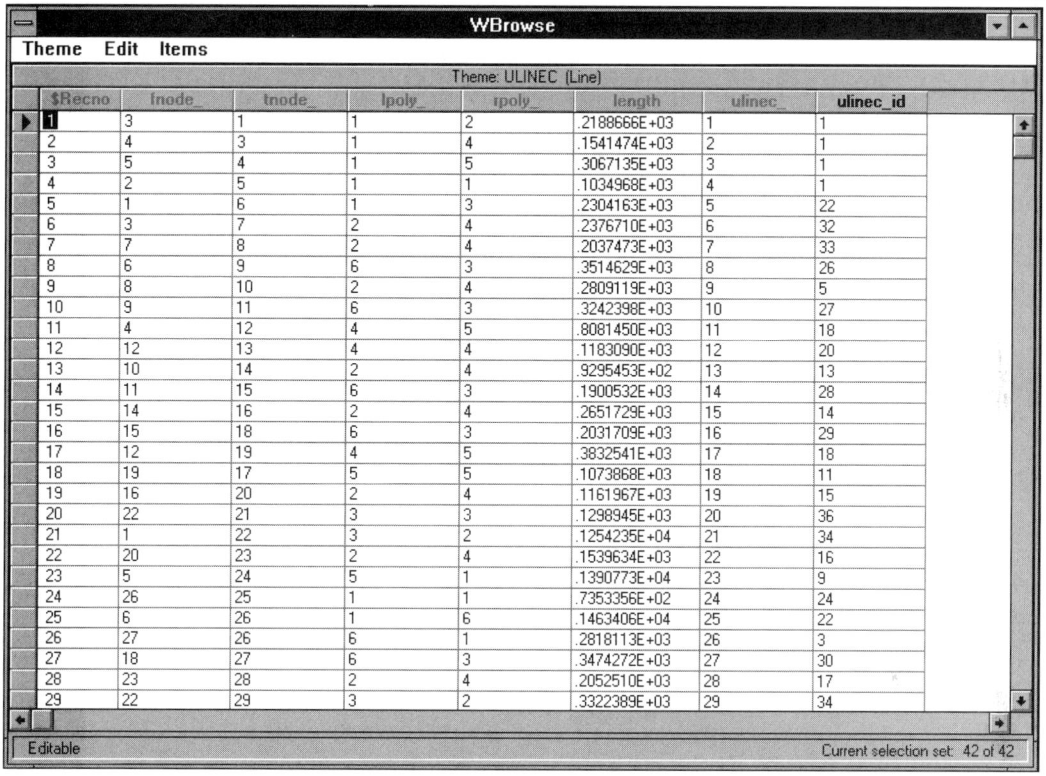

$Recno	fnode_	tnode_	lpoly_	rpoly_	length	ulinec_	ulinec_id
1	3	1	1	2	.2188666E+03	1	1
2	4	3	1	4	.1541474E+03	2	1
3	5	4	1	5	.3067135E+03	3	1
4	2	5	1	1	.1034968E+03	4	1
5	1	6	1	3	.2304163E+03	5	22
6	3	7	2	4	.2376710E+03	6	32
7	7	8	2	4	.2037473E+03	7	33
8	6	9	6	3	.3514629E+03	8	26
9	8	10	2	4	.2809119E+03	9	5
10	9	11	6	3	.3242398E+03	10	27
11	4	12	4	5	.8081450E+03	11	18
12	12	13	4	4	.1183090E+03	12	20
13	10	14	2	4	.9295453E+02	13	13
14	11	15	6	3	.1900532E+03	14	28
15	14	16	2	4	.2651729E+03	15	14
16	15	18	6	3	.2031709E+03	16	29
17	12	19	4	5	.3832541E+03	17	18
18	19	17	5	5	.1073868E+03	18	11
19	16	20	2	4	.1161967E+03	19	15
20	22	21	3	3	.1298945E+03	20	36
21	1	22	3	2	.1254235E+04	21	34
22	20	23	2	4	.1539634E+03	22	16
23	5	24	5	1	.1390773E+04	23	9
24	26	25	1	1	.7353356E+02	24	24
25	6	26	1	6	.1463406E+04	25	22
26	27	26	6	1	.2818113E+03	26	3
27	18	27	6	3	.3474272E+03	27	30
28	23	28	2	4	.2052510E+03	28	17
29	22	29	3	2	.3322389E+03	29	34

The ULINE FAT.

The attributes created include the length of each line segment, the *from* and *to* nodes, the left and right polygons, as well as the internal and User-IDs. For each entity in the drawing file, ArcCAD created a record in the database and a corresponding User-ID. Later, you will learn how to identify each feature's corresponding record in the database and how to add your own data to the database.

Polygon Themes

For a polygon theme, ArcCAD creates a coverage containing a series of lines for each record in the attribute table. To do this ArcCAD uses a "label point," a point inside each polygon boundary, to which it attaches the FAT record. These label points can be created by placing an AutoCAD entity inside each of the polygons and including this entity in the data selection. ArcCAD will use this entity to attach feature attributes. If no entity exists within the polygon as it is written to the theme, the Createlab command can automatically create label points at a later time.

In the next section you will create a polygon theme without label points. Once the theme has been created and cleaned, the Createlab command will be used to create labels for the polygons.

The theme created will contain the soils information for the project.

To begin the process, set the soils layer current and turn off all other layers.

```
Command: la
LAYER ?/Make/Set/New/ON/OFF/Color/Ltype/Freeze/Thaw
  /LOck/Unlock: s
New current layer <WORK>: soils
?/Make/Set/New/ON/OFF/Color/Ltype/Freeze/Thaw/LOck/
  Unlock: of
Layer name(s) to turn Off: *
Really want layer SOILS (the CURRENT layer) off?  <N>
  ENTER
?/Make/Set/New/ON/OFF/Color/Ltype/Freeze/Thaw/LOck/
  Unlock: ENTER
```

To define a theme for the soils, select Defthm from the Themes menu and fill out the dialog box as shown in the following figure.

Defthm dialog box for LSOILS.

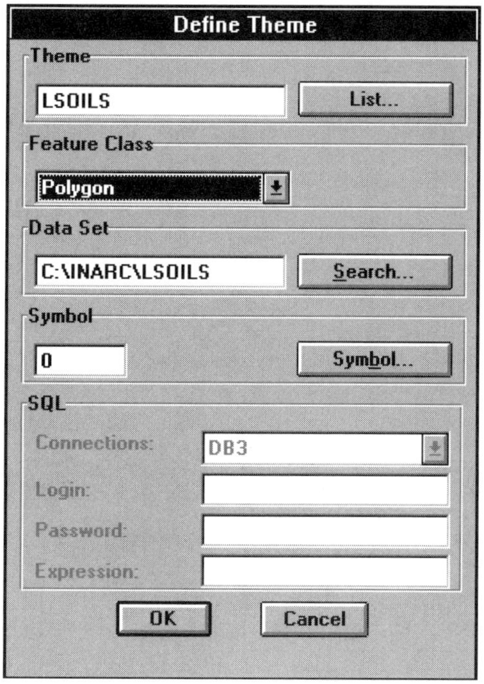

Now execute the Addfeat command from the Features menu. This will write the selected AutoCAD entities to the spatial database.

```
Command: addfeat
Theme name (?/<LSOILS>):ENTER
```

Window all objects on screen

```
Select objects: Other corner: 12 found
Select objects: ENTER
Optional property table (?/<None>):ENTER
20 written to C:\INARC\LSOILS, 0 duplicated, 0
  ignored.
Output theme (?/<LSOILS>):ENTER
```

Select Clean from the dialog box.

```
Dangle length <Default>:ENTER
Fuzzy tolerance <Compute>:ENTER
Cleaning LSOILS.
Sorting...
Intersecting...
Assembling polygons...
```

```
    Sorting input file...
    Sorting label file...
    Processing...
    Assigning final IDs...
    Writing arc file...
    Generating polygon report...
  Creating Polygon attribute table.
  Dropping entity-feature links...
  Unlinking...
```

ArcCAD has now copied the AutoCAD entities into a spatial database. The FAT for this data set can now be viewed using the Wbrowse command found on the Attributes menu.

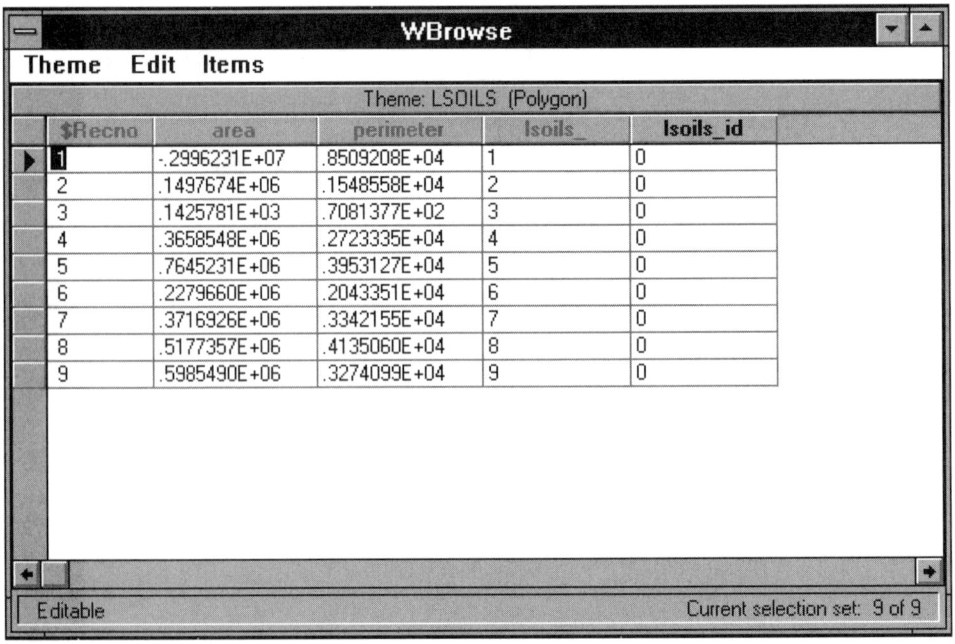

$Recno	area	perimeter	lsoils_	lsoils_id
1	-.2996231E+07	.8509208E+04	1	0
2	.1497674E+06	.1548558E+04	2	0
3	.1425781E+03	.7081377E+02	3	0
4	.3658548E+06	.2723335E+04	4	0
5	.7645231E+06	.3953127E+04	5	0
6	.2279660E+06	.2043351E+04	6	0
7	.3716926E+06	.3342155E+04	7	0
8	.5177357E+06	.4135060E+04	8	0
9	.5985490E+06	.3274099E+04	9	0

Wbrowse of the LSOILS theme.

In the FAT you should notice that ArcCAD did not create User-IDs for the polygons. User-IDs were not created because the polygons did not have label points. The two methods of creating label points for polygons are as follows:

Place AutoCAD points, blocks, or text entities within the polygons before using Addfeat.

or

Use the ArcCAD Createlab command once a polygon coverage has been created.

If an entity exists within the polygon and is included in the Addfeat selection set, ArcCAD will use the insertion point of this entity as the label point. If you use the Createlab command, the User-IDs will be created in an arbitrary order, with each polygon getting a unique number.

Since the soils polygons did not have label points, run Createlab to add them to the FAT.

From the Topology menu choose Createlab.

```
Command: createlab
Theme name (?/<LSOILS>):ENTER
User-ID base <0>:ENTER
```

The Createlab command uses a User-ID base value to sequentially add a unique ID to each polygon. The IDs, however, are not immediately written to the FAT. In order to combine the new IDs with the FAT, the theme must be built.

To build the LSOILS theme, choose Build from the Topology menu.

```
Command: build
Input theme (?/<LSOILS>): ENTER
Output theme (?/<LSOILS>):ENTER
Building polygons...
Processing...
  Sorting input file...
  Sorting label file...
  Processing...
  Assigning final IDs...
  Writing arc file...
  Generating polygon report...
Creating Polygon attribute table.
Dropping entity-feature links...
```

The Build command added the new IDs to the FAT. The file can be viewed using the Wbrowse command on the Attributes menu. The new table should look like the following figure.

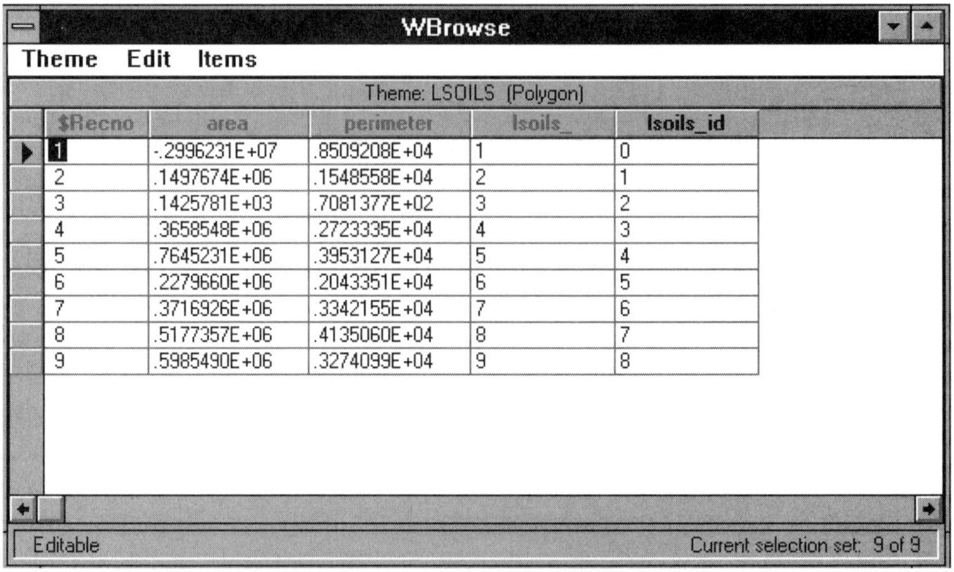

$Recno	area	perimeter	lsoils_	lsoils_id
1	-.2996231E+07	.8509208E+04	1	0
2	.1497674E+06	.1548558E+04	2	1
3	.1425781E+03	.7081377E+02	3	2
4	.3658548E+06	.2723335E+04	4	3
5	.7645231E+06	.3953127E+04	5	4
6	.2279660E+06	.2043351E+04	6	5
7	.3716926E+06	.3342155E+04	7	6
8	.5177357E+06	.4135060E+04	8	7
9	.5985490E+06	.3274099E+04	9	8

Wbrowse of the LSOILS theme with User-IDs.

Point Themes

Point themes are ArcCAD's simplest theme type. A point theme is used to describe any single point feature. What is described by points is directly related to the scale of the theme you are creating. For example, on a map of the United States, points may be used to map the location of cities. On a map of a particular city, points may be used to locate manholes or utility poles. A map of a particular site within the city may use points to locate wells or trees. In the project data you will create two point themes. The first will hold the fire hydrant locations and the second will hold the street light locations.

Point themes are never cleaned. Once created, they only need to be built to create topology. The table on page 59 shows which AutoCAD entities can be used to create point themes. In the project data, the themes will be created from AutoCAD blocks.

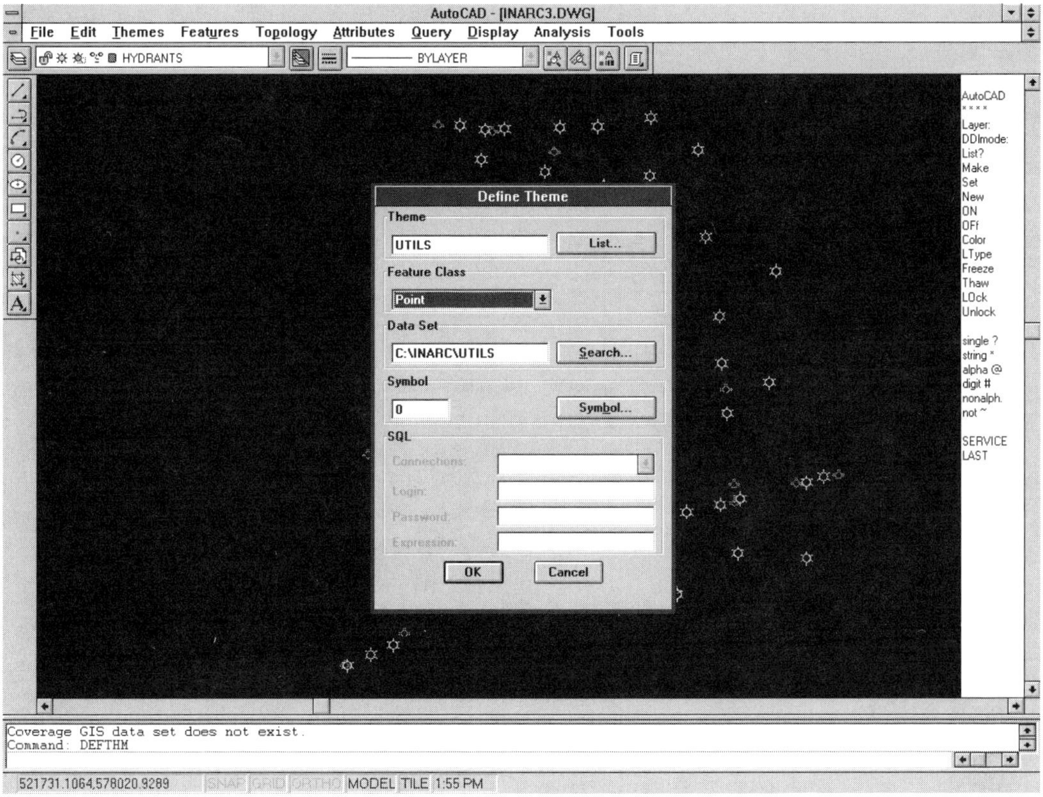

Defthm dialog box for UTILS.

Begin creating the utility location themes by setting the Hydrants layer current and turning off all other layers, except Lights.

```
Command: LA
LAYER ?/Make/Set/New/ON/OFF/Color/Ltype/Freeze/Thaw/
  LOck/Unlock: S
New current layer <SOILS>: HYDRANTS
?/Make/Set/New/ON/OFF/Color/Ltype/Freeze/Thaw/LOck/
  Unlock: OF
Layer name(s) to turn Off: *
Really want layer HYDRANTS (the CURRENT layer) off?
  <N>: ENTER
```

```
?/Make/Set/New/ON/OFF/Color/Ltype/Freeze/Thaw/LOck/
  Unlock: ON
Layer name(s) to turn On: LIGHTS
?/Make/Set/New/ON/OFF/Color/Ltype/Freeze/Thaw/LOck/
  Unlock: ENTER
```

With the fire hydrants and lights showing, define a point theme using the Defthm command found on the Themes menu. The dialog box should be filled out as follows.

Optional Property Tables

Optional property tables are used to extract AutoCAD properties from entities. These properties can then be added to the theme's FAT. In the case of the utilities theme, the name of the AutoCAD block will be extracted. This will identify whether the utility is a fire hydrant or a light pole. Later, the theme can be queried to display only light poles or only fire hydrants.

ArcCAD adds AutoCAD properties to a FAT by using a record theme defining the properties to extract. For example, to extract the block name from the entity, an item would be added to the record theme with the name ACAD_BLOCK. To extract the layer name of the entity, an item named ACAD_LAYER would be added.

To automate the creation of a coverage with attributes, you will define a record theme for the Utilities' properties. To create the optional property table, select Defthm from the Themes menu and fill out the dialog box as shown in the following figure.

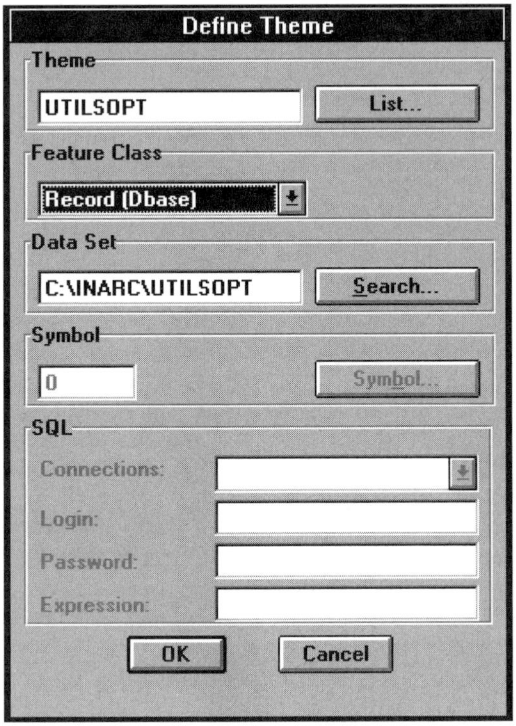

Defthm dialog box for UTILSOPT.

The next step is to add an item to the table that is common with the Point theme. Remember, when creating any theme, ArcCAD creates a default set of items. The item that is used to link spatial themes to their property tables is the User-ID, where User is the name of the coverage being created. For this reason, the first item you will add to the property table is Utils_ID.

Follow the command process below to add this item to the UTILSOPT theme.

```
Command: ADDITEM
Theme name (?/<UTILSOPT>):ENTER
Item name (?): UTILS_ID
Item width: 11
Item type?  (Character/Date/<Numeric>):ENTER
Item decimal places <0>:ENTER
Start item <Last>:ENTER
```

NOTE: *The ID field of the point theme is added to the optional property table to provide a common item for a Relate or Join later on.*

To extract the block name from the UTILS theme, you will need to add an item name ACAD_BLOCK to the UTILSOPT record theme.

```
Command: ADDITEM
Theme name (?/<UTILSOPT>): ENTER
Item name (?): ACAD_BLOCK
Item width: 10
Item type?  (Character/Date/<Numeric>): C
Start item <Last>: ENTER
```

NOTE: *The ACAD_BLOCK item tells ArcCAD to extract the AutoCAD block name from each entity to the optional property table.*

Now that the themes have been defined to hold the points and the AutoCAD properties, the features can be added.

```
Command: ADDFEAT
Theme name (?/<UTILSOPT>): UTILS
Select objects: Other corner: 112 found
Select objects: ENTER
Optional property table (?/<None>): UTILSOPT
112 written to C:\INARC\UTILS, 0 duplicated, 0
  ignored.
Output theme (?/<UTILS>): ENTER
```

Select Build Now from the dialog box.

```
Building points...
Creating Point attribute table.
Dropping entity-feature links...
Unlinking...
```

The Addfeat and Build commands have created a coverage from the AutoCAD blocks and added the properties of these blocks to a database file. Use the Wbrowse command on the Attributes menu to view the attribute table of the UTILS POINT theme and the UTILSOPT record theme.

```
Command: wbrowse
Theme name (?/<UTILS>): ENTER
```

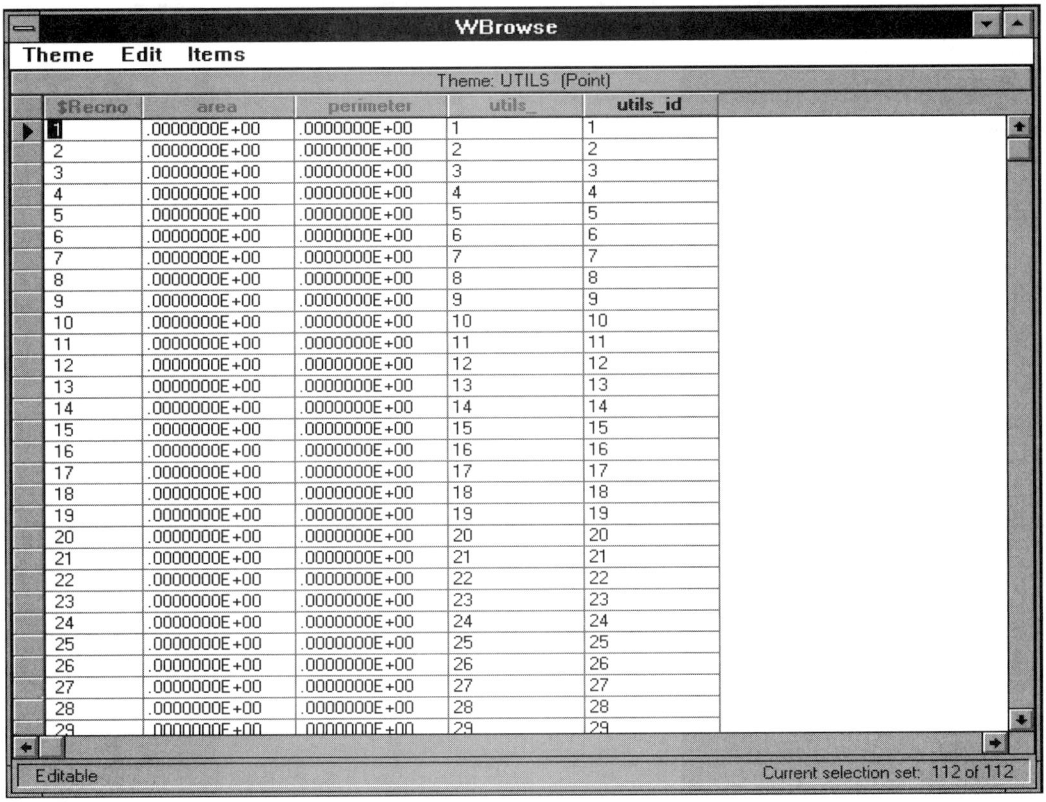

WBrowse

Theme Edit Items

Theme: UTILS (Point)

$Recno	area	perimeter	utils_	utils_id
1	.0000000E+00	.0000000E+00	1	1
2	.0000000E+00	.0000000E+00	2	2
3	.0000000E+00	.0000000E+00	3	3
4	.0000000E+00	.0000000E+00	4	4
5	.0000000E+00	.0000000E+00	5	5
6	.0000000E+00	.0000000E+00	6	6
7	.0000000E+00	.0000000E+00	7	7
8	.0000000E+00	.0000000E+00	8	8
9	.0000000E+00	.0000000E+00	9	9
10	.0000000E+00	.0000000E+00	10	10
11	.0000000E+00	.0000000E+00	11	11
12	.0000000E+00	.0000000E+00	12	12
13	.0000000E+00	.0000000E+00	13	13
14	.0000000E+00	.0000000E+00	14	14
15	.0000000E+00	.0000000E+00	15	15
16	.0000000E+00	.0000000E+00	16	16
17	.0000000E+00	.0000000E+00	17	17
18	.0000000E+00	.0000000E+00	18	18
19	.0000000E+00	.0000000E+00	19	19
20	.0000000E+00	.0000000E+00	20	20
21	.0000000E+00	.0000000E+00	21	21
22	.0000000E+00	.0000000E+00	22	22
23	.0000000E+00	.0000000E+00	23	23
24	.0000000E+00	.0000000E+00	24	24
25	.0000000E+00	.0000000E+00	25	25
26	.0000000E+00	.0000000E+00	26	26
27	.0000000E+00	.0000000E+00	27	27
28	.0000000E+00	.0000000E+00	28	28
29	.0000000E+00	.0000000E+00	29	29

Editable Current selection set: 112 of 112

UTILS POINT attribute table.

```
Command: wbrowse
Theme name (?/<UTILS>): utilsopt
```

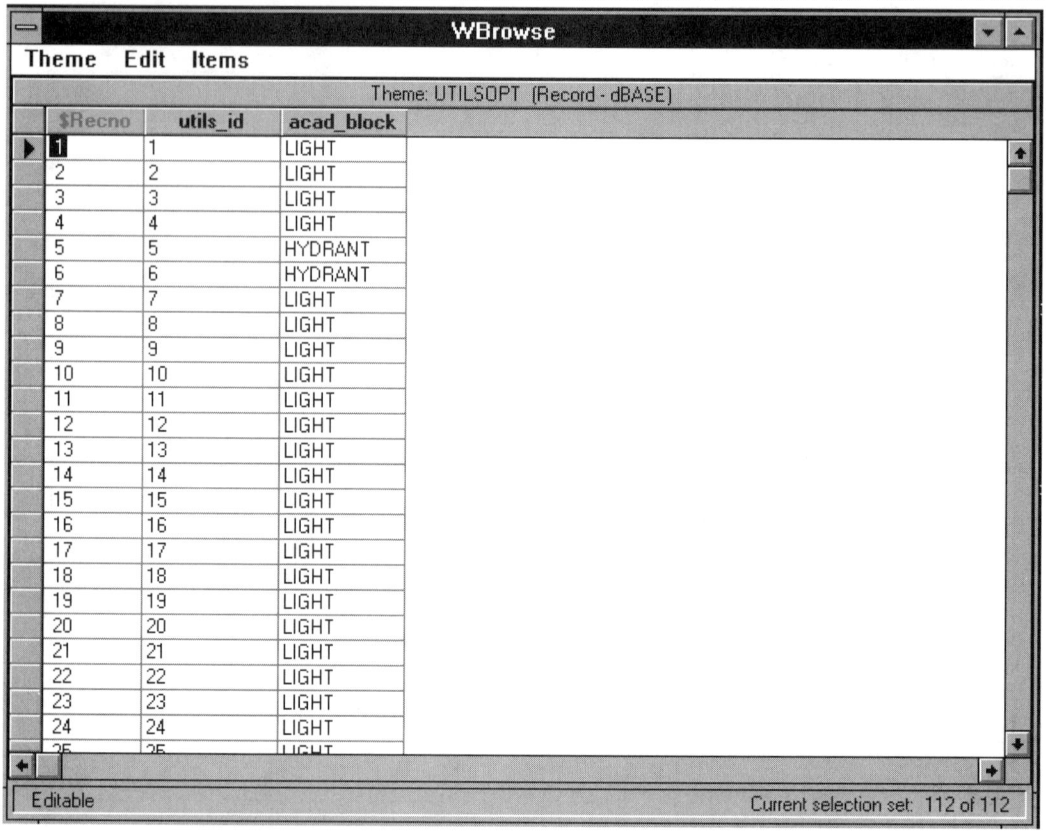

$Recno	utils_id	acad_block
1	1	LIGHT
2	2	LIGHT
3	3	LIGHT
4	4	LIGHT
5	5	HYDRANT
6	6	HYDRANT
7	7	LIGHT
8	8	LIGHT
9	9	LIGHT
10	10	LIGHT
11	11	LIGHT
12	12	LIGHT
13	13	LIGHT
14	14	LIGHT
15	15	LIGHT
16	16	LIGHT
17	17	LIGHT
18	18	LIGHT
19	19	LIGHT
20	20	LIGHT
21	21	LIGHT
22	22	LIGHT
23	23	LIGHT
24	24	LIGHT
25	25	LIGHT

UTILS optional property table.

Later, you will temporarily relate and permanently merge the attribute table with the optional property table.

You have created your first theme using an optional property table. Now let's look at a simpler method of extracting AutoCAD properties to a FAT.

The Ddfeat Command

You will use the Ddfeat command to define a theme, set the AutoCAD properties to extract, and add features to the theme, all in one command.

To begin, set the trees layer current and turn off all other layers. The drawing should look as shown in the following figure.

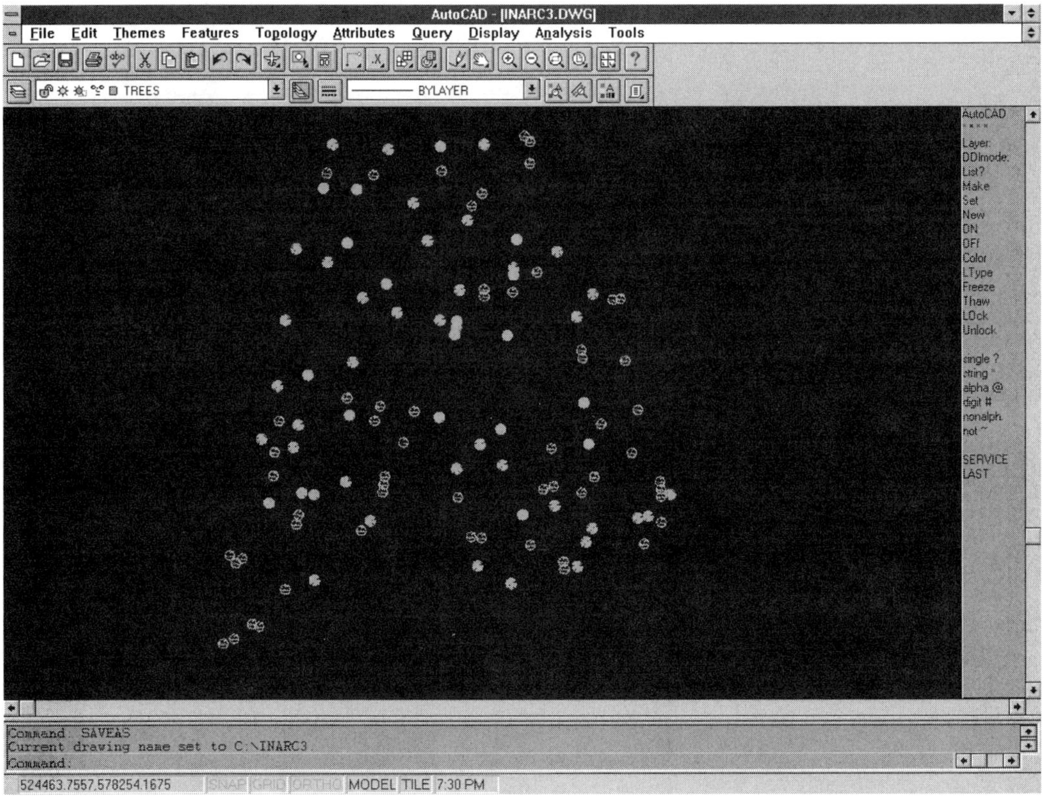

The drawing with trees displayed.

Next, select Dddfeat from the Features menu. The Ddfeat dialog box is shown below.

The Ddfeat dialog box.

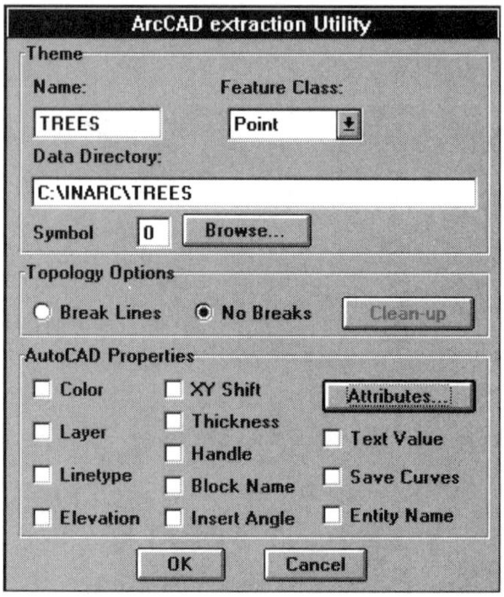

Fill out the dialog box as shown. To extract the block attributes associated with the tree symbols, select the Attributes button on the Ddfeat dialog box. The following subdialog box is displayed. Select each of the attributes and set the item type to Integer and length to 8 as shown below.

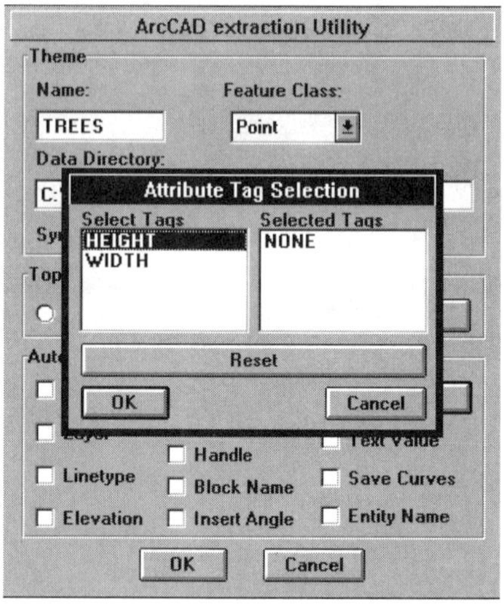

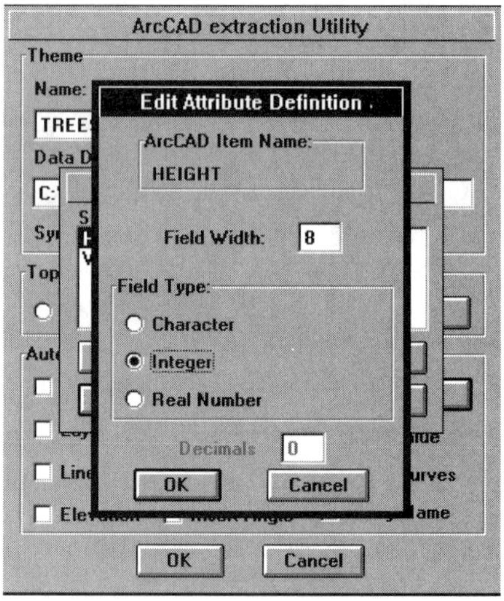

LEFT: The Attribute Tag Selection dialog box. RIGHT: The Attribute Definition dialog box.

Once the dialog box is complete, select OK. On the Extraction Utility dialog box, toggle on the Block Name check box and select O.K.

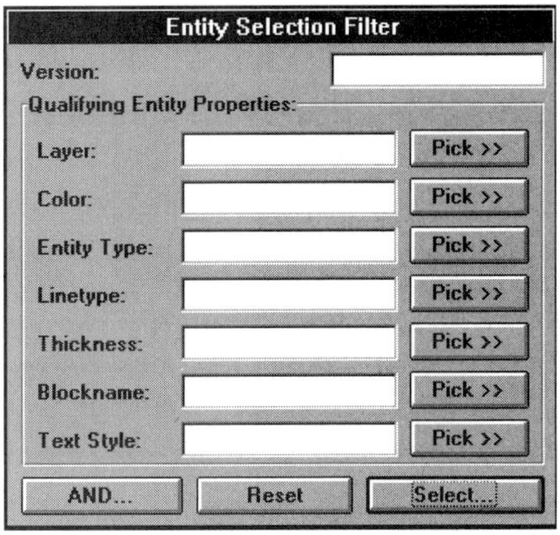

The Entity Selection Filters dialog box.

Choose Select from the Entity Selection Filters dialog box. At the Select Objects prompt, Window the entire drawing. At the Update Topology dialog box, select Build Now. ArcCAD has created a new theme, written the trees to it, and added the block attributes and block name to the FAT. To view the FAT for the TREES theme, execute the Wbrowse command. At the theme name prompt, enter TREES.

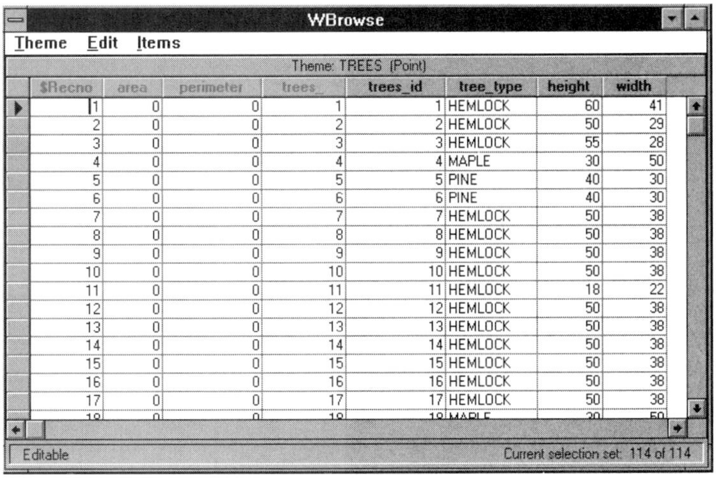

WBrowse

Theme Edit Items

Theme: TREES (Point)

$Recno	area	perimeter	trees_	trees_id	tree_type	height	width
1	0	0	1	1	HEMLOCK	60	41
2	0	0	2	2	HEMLOCK	50	29
3	0	0	3	3	HEMLOCK	55	28
4	0	0	4	4	MAPLE	30	50
5	0	0	5	5	PINE	40	30
6	0	0	6	6	PINE	40	30
7	0	0	7	7	HEMLOCK	50	38
8	0	0	8	8	HEMLOCK	50	38
9	0	0	9	9	HEMLOCK	50	38
10	0	0	10	10	HEMLOCK	50	38
11	0	0	11	11	HEMLOCK	18	22
12	0	0	12	12	HEMLOCK	50	38
13	0	0	13	13	HEMLOCK	50	38
14	0	0	14	14	HEMLOCK	50	38
15	0	0	15	15	HEMLOCK	50	38
16	0	0	16	16	HEMLOCK	50	38
17	0	0	17	17	HEMLOCK	50	38

Editable — Current selection set: 114 of 114

The TREES FAT.

To exit the Wbrowse command, choose EXIT from the dialog box's Theme menu.

DXF Files

Drawing Interchange Files (DXF) are ASCII files written and read by most CAD systems. With AutoCAD's ability to read in DXF files, essentially all CAD data is available for use in ArcCAD. To bring a DXF file into AutoCAD, execute the `DXFIN` command at the command prompt.

> **NOTE:** *The DXFIN command must be used in a blank AutoCAD drawing. If any data has been added to the drawing, the DXF file will not import correctly. To ensure that a drawing file is in fact blank, start a new drawing without a prototype by specifying "No prototype" in the dialog box.*

Once the DXF file has been imported, the file is the same as any AutoCAD drawing and entities can be used as source data for ArcCAD.

Add the land use information to your project from an existing file. First, import the DXF file into a blank AutoCAD drawing.

Use the ESRI_PREFS command to toggle "Start ArcCAD when AutoCAD starts" Off.

Type `new` at the command prompt and save the changes to the current drawing. Enter `Landuse` as the new drawing name and toggle No Prototype On.

Create New Drawing dialog box.

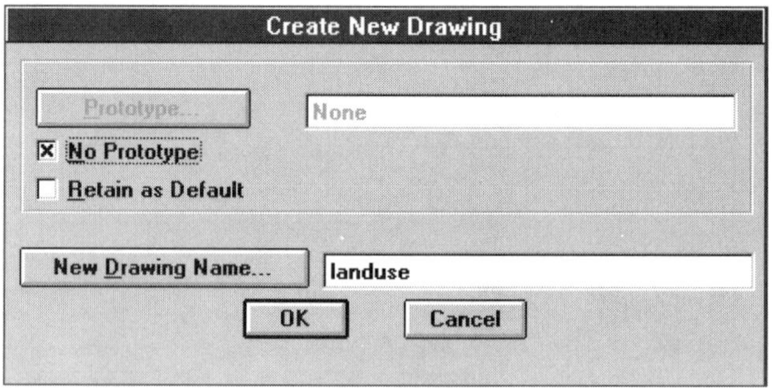

At the command prompt, enter `DXFIN` and select LANDUSE.DXF as the file to import.

Select DXF File dialog box.

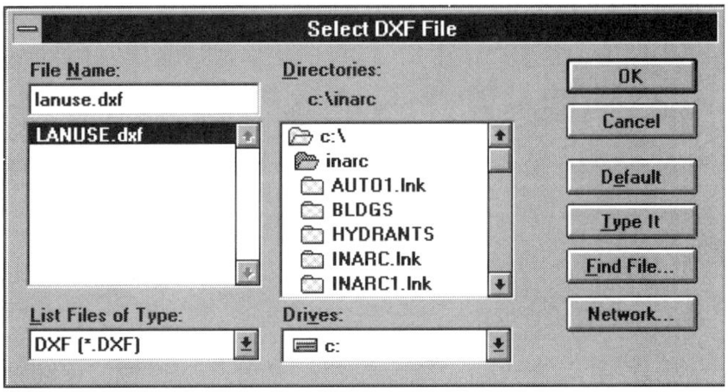

When the command has finished, do a Zoom Extents. The Landuse polygons should look like the following image.

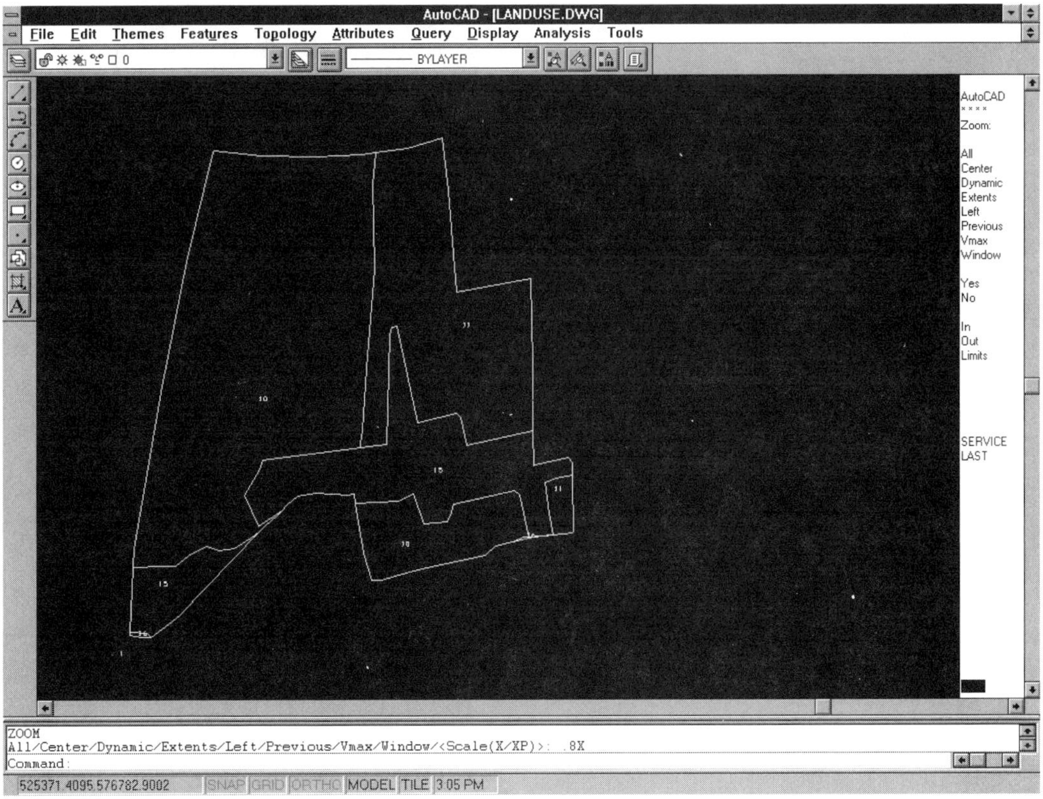

Landuse polygons.

To complete the process of creating the landuse polygon coverage, a theme must be defined, features added, and topology built. The Ddfeat command will be used again here to accomplish all three of these steps.

Select Ddfeat from the Features menu.

```
Command: DDFEAT
```

The Ddfeat dialog box.

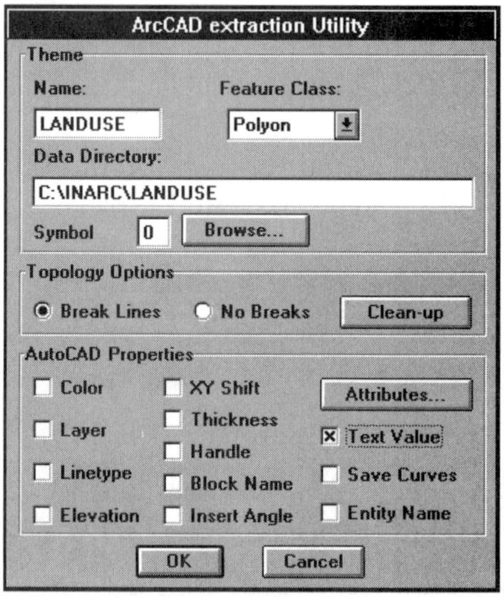

Fill out the dialog box as shown in the previous figure and select OK. Selecting OK brings you to the Attribute Definition dialog box.

TheEdit Attribute Definition dialog box.

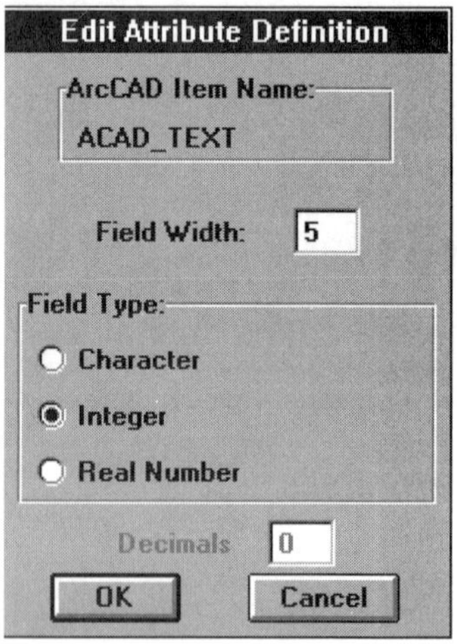

This dialog box prompts for the size and type of fields to create to hold the AutoCAD properties being extracted. In this case you are extracting the AutoCAD text value.

Set the size to 5 and the type to integer. Selecting OK on this dialog box brings you to the last of the Ddfeat dialog boxes.

The Ddfeat Entity Selection dialog box.

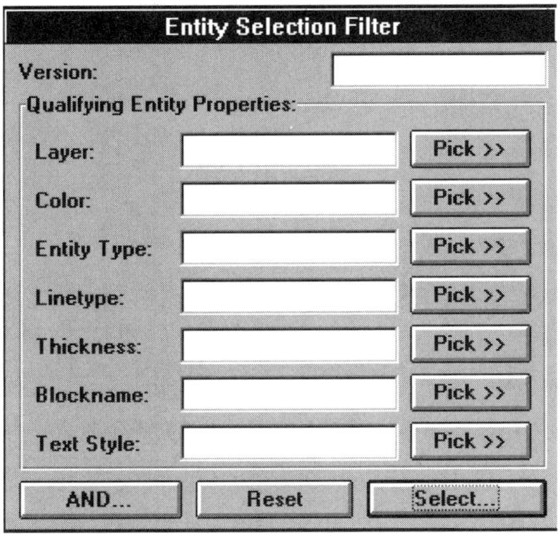

This dialog box works as a selection filter. It allows you to choose all of the objects on a specific layer or of a specific text style. Choose Select from the bottom of this dialog box, window all objects on screen, and press Enter to complete the command.

You have now imported a DXF file and created a new coverage based on its features. To continue with this chapter, return to the INARC3 drawing file.

ARC/INFO Export Files

Although ARC/INFO can create DXF files, there is a better way to transfer data from ARC/INFO to ArcCAD. ARC/INFO can transfer its coverages to "export files" readable by ArcCAD. Export files allow the transfer of spatial information (the graphics), as well as the feature attributes (the database).

The ArcCAD Import command is used to convert these export files to ArcCAD coverages. Themes are then defined, and the data may be queried, analyzed, and displayed.

You will now import an ARC/INFO export file representing the Buildings information for our project.

> **NOTE:** *ArcCAD versions prior to 11.4 can only read single precision export files. If you are using ArcCAD 11.3 or prior, be sure to notify the ARC/INFO operator that the export files need to be single precision.*

The command for reading export files is ESRI_IMPORT and can be found on the Tools pulldown menu under Translate. The input for the ESRI_IMPORT command is shown below.

```
Command: ESRI_IMPORT
Type of Import? (Datafile/Textfile/<Coverage>):ENTER
Interchange file name (?): BLDGS
Coverage name: BLDGS
Use an xy shift? (Yes/No/<Calculate>):ENTER
Note: Single precision interchange file. No shift
  will occur!
```

ArcCAD has now created a coverage from the export file. In order to access this coverage, a theme must now be defined. This theme will allow AutoCAD to access the coverage information.

Execute the Defthm command on the Themes menu and fill out the dialog box as shown below.

Defthm for Bldgs coverage.

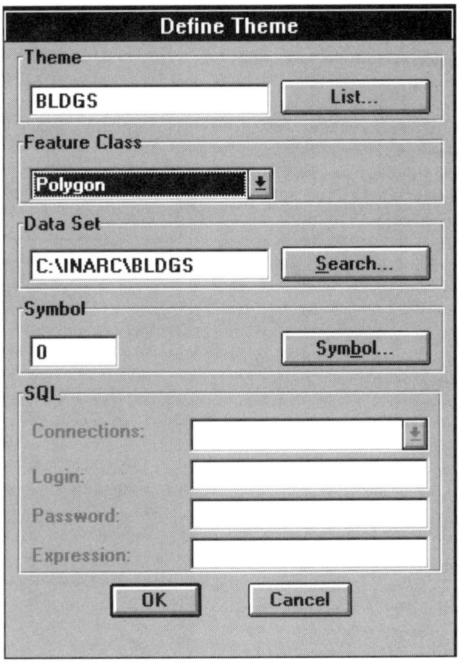

The themes data may now be reviewed using the Wbrowse command found on the Attributes menu.

PC ARC/INFO Coverages

PC ARC/INFO coverages are identical to ArcCAD coverages. To use a PC ARC/INFO coverage in ArcCAD, you simply define a theme identifying the coverage to AutoCAD. All spatial and attribute data is accessible from a PC ARC/INFO coverage.

There are two coverages supplied with the sample data set. These coverages represent the street centerlines and parcel information. To access these coverages, you must define themes using the Defthm command and point to the data sets. The input for the theme definitions is provided below.

Defthm for Streets.

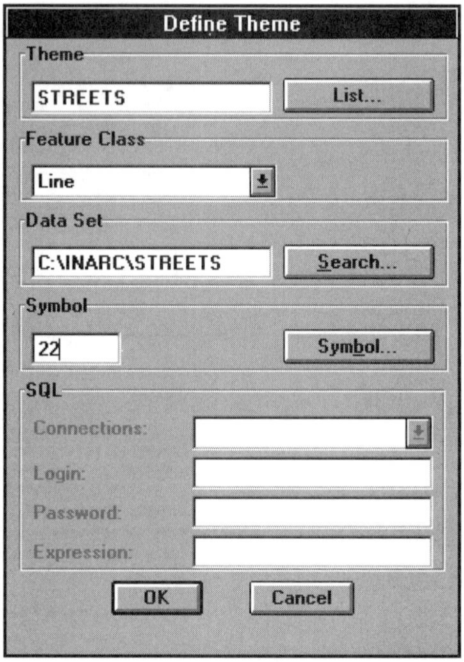

Defthm for Parcels.

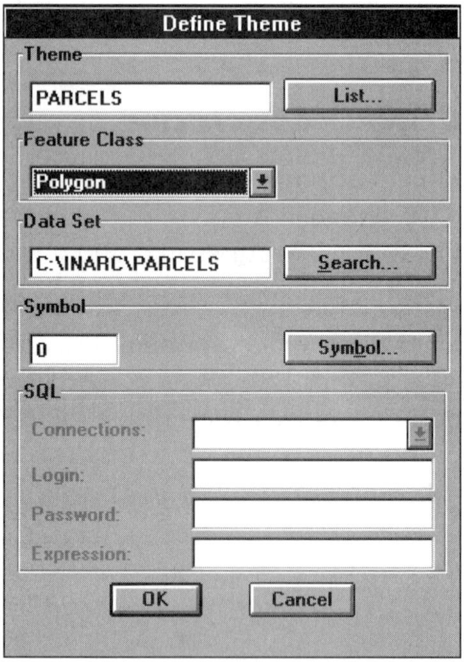

The Street centerline and Parcels can now be displayed in the drawing using the Arclines and Polys commands, respectively. Follow the input below to display this information.

```
Command: LA
LAYER ?/Make/Set/New/ON/OFF/Color/Ltype/Freeze/Thaw/
  LOck/Unlock: M
New current layer <HYDRANTS>: STREETS
?/Make/Set/New/ON/OFF/Color/Ltype/Freeze/Thaw/LOck/
  Unlock: ENTER

Command: ARCLINES
Theme name (?/<PARCELS>): STREETS
Display using? (Properties/Item/<Symbol>): ENTER
Line symbol number (0-100) (?/<0>): 1

Command: LA
LAYER ?/Make/Set/New/ON/OFF/Color/Ltype/Freeze/Thaw
  /LOck/Unlock: M
New current layer <STREETS>: PARCELS
?/Make/Set/New/ON/OFF/Color/Ltype/Freeze/Thaw/LOck/
  Unlock: ENTER

Command: POLYS
Theme name (?/<STREETS>): PARCELS
Line symbol number (0-100) (?/<0>): 3
```

Turn all layers on and Zoom Extents. The drawing file should now look like the following figure.

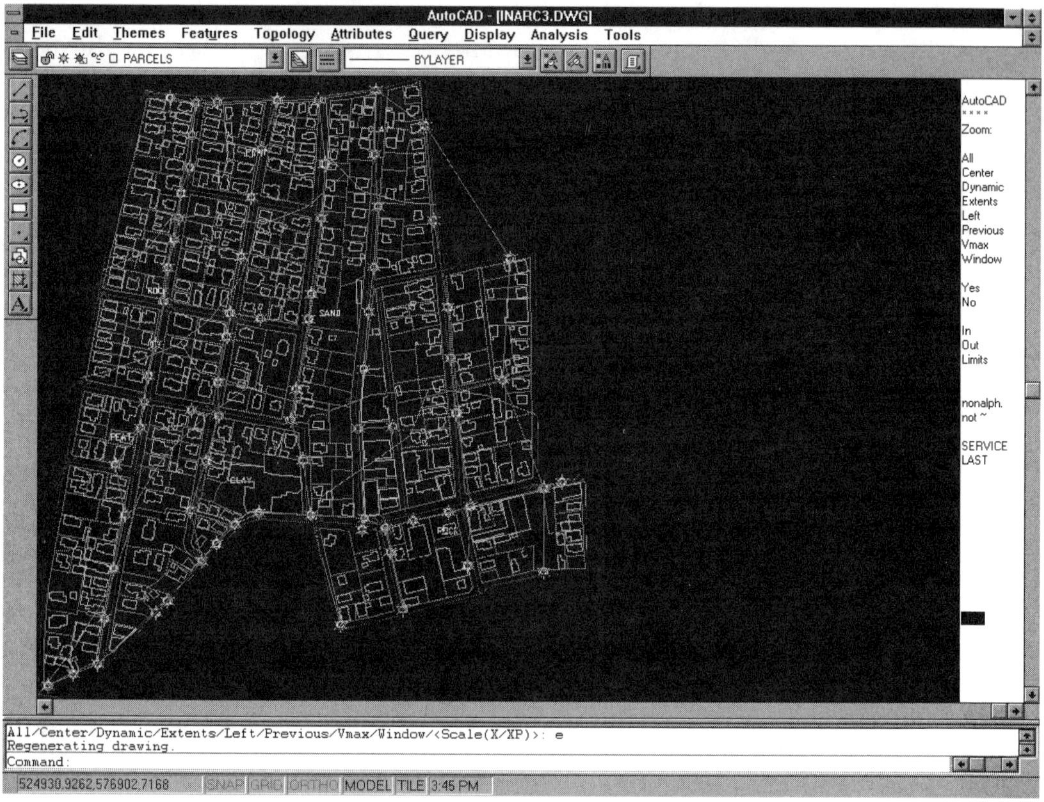

INARC3 drawing file.

Review

In this chapter we have reviewed sources of data for ArcCAD. Although the examples used in this section are the major sources, there are many more types of data that can be translated into AutoCAD or ArcCAD format. These data types include Digital Line Graph (DLG) data and Tiger files. For more information on sources of data, refer to the ArcData data publishing program from ESRI.

You have seen how to create ArcCAD coverages from AutoCAD drawings and DXF files, and how to access coverages created with ARC/INFO and PC ARC/INFO. You should now be able to:

- ❐ Define themes
- ❐ Create optional property tables
- ❐ Build coverages
- ❐ Clean coverages

In the next chapter you will use ArcCAD commands to find possible errors in the data set and AutoCAD commands to correct these errors.

Chapter

Correcting Spatial Data

When Clean Won't Do It

This chapter covers the commands necessary to correct and modify spatial data sets. These commands will allow you to modify point, line, and polygon coverages.

Any discussion of spatial feature modification should first explain entity feature links. These links are the difference between AutoCAD entities and ArcCAD features. A line in an AutoCAD drawing could be either. If a line was created in AutoCAD by displaying information from an ArcCAD theme, ArcCAD will have created a link between the feature attribute table and the entity. An ArcCAD feature consists of the AutoCAD entity, the record in the feature attribute table, and the link between the entity and the table.

When modifying ArcCAD spatial data, you must modify features first and then update the information in ArcCAD. This process of modifying features and updating their themes is covered in this chapter.

Setup

To prepare for this chapter,

> Load ArcCAD and create a new drawing named INARC4.DWG. Load ArcCAD if it is not set to load automatically. Define a polygon theme for the LANDUSE coverage. The dialog box is shown below.

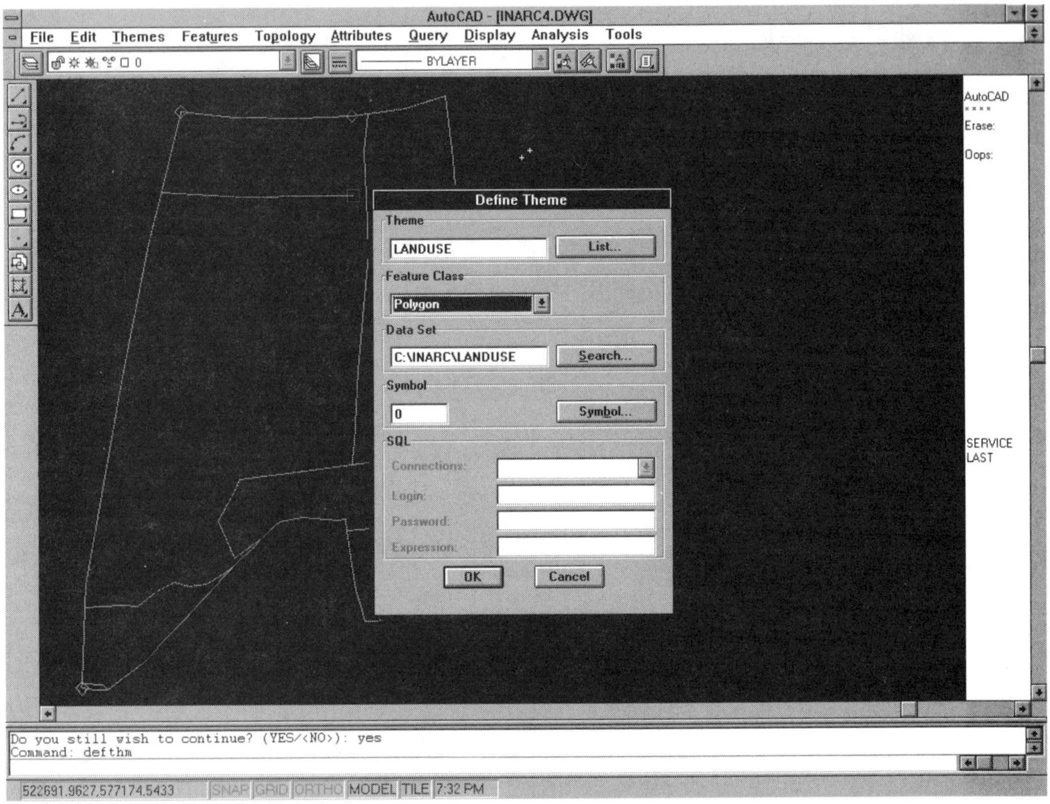

Defthm dialog box for LANDUSE polygon theme.

Define a second theme named LANDUSEL. This theme will be a line theme pointing to the polygon data set. The second theme allows access to the polygon boundaries. The only way to modify polygons is to access the line information. The following dialog box shows the input for this theme definition.

Defthm dialog box for LANDUSEL line theme.

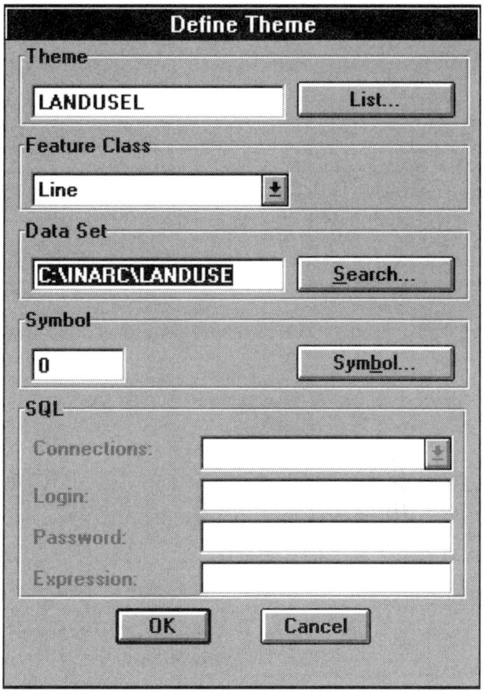

NOTE: *Notice that the data set name for the LANDUSEL theme is Landuse and the feature class is line.*

Execute the Setext command found on the Display menu. At the theme name prompt, enter LANDUSE and press Enter. The Setext command sets the extents of the drawing to the limits of the LANDUSE theme.

Create a new layer named LANDUSE and set it current.

Identifying Polygon Errors

ArcCAD provides several commands on the Topology menu for identifying errors or potential errors in your spatial data. These commands include the following:

- Laberror
- Noderror
- Clrerror
- Resize

These commands allow the identification of errors such as polygons without label points, lines that appear to be dangles, and nodes that exist with no apparent purpose. The next section contains a discussion of each of these errors and how ArcCAD identifies them.

Node Errors

Potential node errors occur when arc endpoints do not connect where expected or when lines overshoot each other (dangles). When ArcCAD evaluates themes for node errors it returns one of two kinds: *dangle nodes* or *pseudo nodes*. *Dangle nodes* occur where a line comes to an end. *Pseudo nodes* occur when ArcCAD detects a node on a line that is not an intersection or endpoint. These dangle and pseudo nodes may or may not be actual errors.

To display the node errors for the LANDUSE polygons, bring the polygon boundaries into the drawing using the Arclines command found on the Display menu under Lines.

```
Command: ARCLINES
Theme name (?/<LANDUSEL>): ENTER
Line symbol number (0-100) (?/<0>): 5
```

With the lines now in the drawing, select the Noderror command found on the Topology menu.

```
Command: NODERROR
Theme name (?/<LANDUSE>): LANDUSEL
Output report to: ENTER
Error type? (Dangle/Pseudo/<Both>): ENTER
```

LANDUSE polygons with node errors.

In the previous figure the dangle nodes are shown as red squares and the pseudo nodes as yellow diamonds. Being able to display potential errors in this way allows you to evaluate a large area quickly.

You should now zoom into each node to correct the problems.

As you zoom, notice that the error labels are sometimes too large to accurately identify the problem area. In these situations the Resize command found on the Topology menu can be used to update the size of the error labels based on the current screen size.

Test this by zooming into the dangle node on the lower right corner of the northwest landuse polygon. Execute the Resize command from the Topology menu. All error labels are automatically resized based on the current screen size.

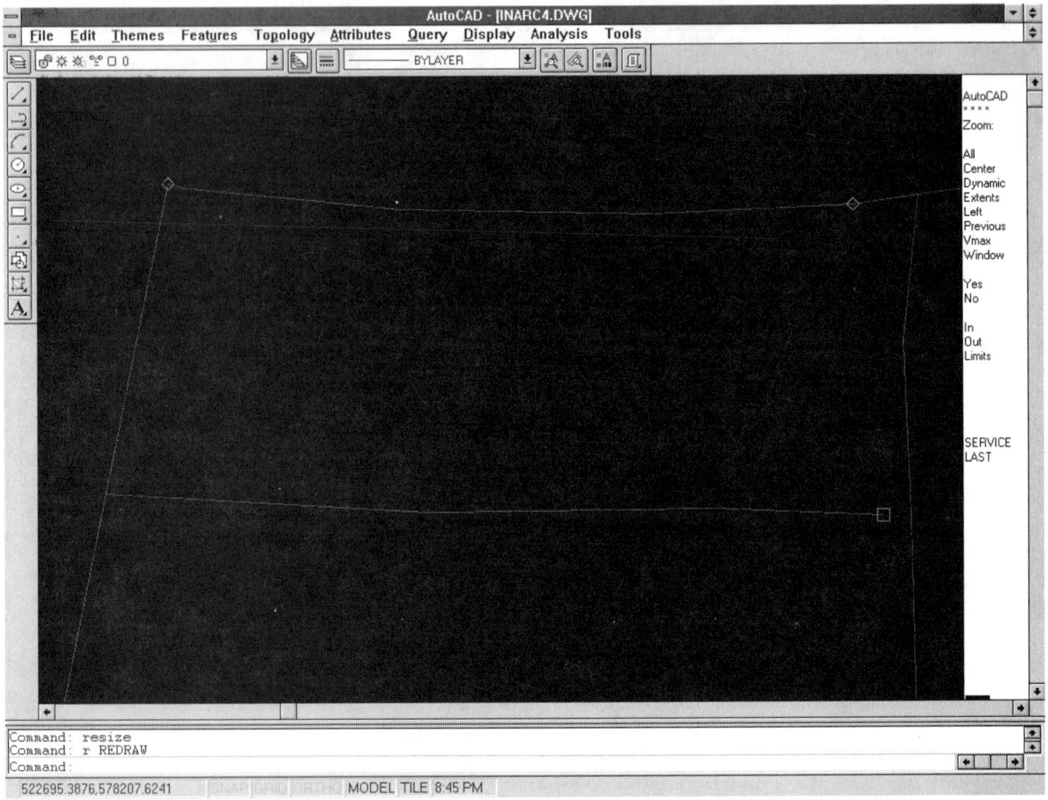

LANDUSE polygons with node errors.

Once the errors have been corrected, the Clrerror command can be used to remove the error labels from the drawing. The following sections show you how to correct these errors in the drawing and update the coverage.

Label Errors

You learned that a polygon is defined by a series of lines bounding a label point. You also know that this label point is used by ArcCAD to link records in the feature attribute table. Therefore, a polygon without a label point will have no record in the feature attribute table. The Laberror command on the Topology menu can be used to display a blue star in any polygon that lacks a label point.

Label errors typically happen when a polygon theme is created using AutoCAD entities as label points and one polygon is missing an entity. The other time label point errors typically happen is when a theme is created and the Createlab command is not used to add IDs to the coverage.

In the LANDUSE theme, the polygon at the southeast corner of the city has no label point. To display this error, execute the Laberror command from the Topology menu. Enter LANDUSE at the Theme name prompt..

The Laberror command plots two symbols on the drawing. A white star roughly in the center of the city and a blue star in the southeast polygon. The white star identifies the universe polygon. The blue star identifies a polygon that is missing a label point.

Landuse polygons with label errors.

The Universe Polygon

The universe polygon is an imaginary polygon that has an internal ID of 1, a User-ID that must always equal 0, and an area that is the inverse of the sum of all of the polygons in the theme. It is very important that the universe polygon always has a User-ID of 0. If the User-ID is anything other than 0, the universe polygon has more than likely been assigned a label point in error and problems will occur when querying and analyzing the data set.

The theory behind the universe polygon is that the inverse of the sum of all of the area contained within the polygons equals all of the area that is *not* within the polygons. With that information, ArcCAD can use the universe polygon to identify which points are outside the polygon coverage.

You have seen how to identify errors in polygon coverages. Let's look at how to correct these areas and update the topology.

Modifying Polygons

You will begin your session on correcting spatial errors by working with a complex example. Previously, you learned that polygon coverages are stored in ArcCAD as a label point bounded by a series of lines. In order to edit the polygons, the lines creating the polygon must be accessed and displayed in the AutoCAD drawing. To access these lines, a line theme must be defined. The line theme will use the polygon data set as its source.

When Arclines is used to display the theme, ArcCAD will draw the polygon boundaries, but rather than create a link to the area (label point), ArcCAD will create a link to the lines bounding the label point. The lines can now be modified, and the polygon coverage can be rebuilt or cleaned.

Since the line theme and the polygon theme share one data set, when one theme is cleaned or built, the topology for the other theme is also updated. For example, a line and polygon theme have already been created for the LANDUSE polygons. If the LANDUSE theme is modified and cleaned, the data set referenced by the LANDUSEL line theme is also updated.

Correcting Node Errors

You will begin this section by correcting the node errors identified earlier.

To begin, erase all entities from the drawing using the AutoCAD Erase command. Set the text size to 30, and display the LANDUSE polygon boundaries in the drawing using the Arcs command found on the Display menu under Lines.

```
Command: TEXTSIZE
New value for TEXTSIZE <0.2000>: 30
Command: ARCS
Theme name (?/<LANDUSE>): LANDUSEL
Display IDs? (Only/Yes/<No>): YES
```

NOTE: *Note that the Arcs command was used on the LANDUSEL theme. The theme accesses the LANDUSE boundary lines, rather than the polygon area.*

The LANDUSE boundaries are shown in the following figure.

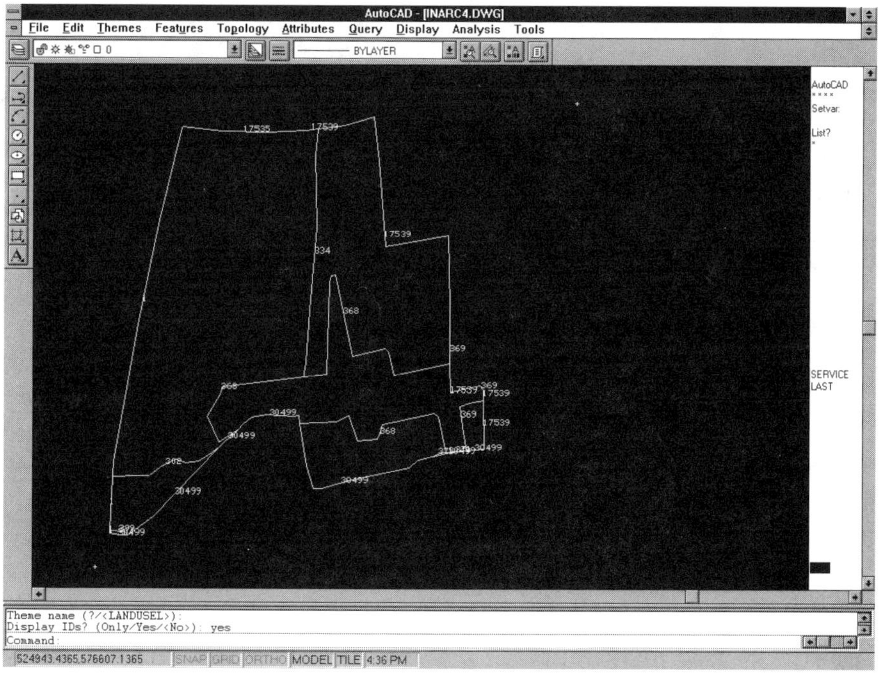

LANDUSE boundary lines.

The Arclines command has drawn the LANDUSE boundary arcs into the drawing and created a link between them and the feature attribute table. The link can be seen here. Each line has been labeled with its User-ID from the feature attribute table. Errors in the boundaries can now be corrected, and the Modfeat command can be used to write the changes to the coverage. Once Modfeat has been used to update the spatial data, the coverage can be cleaned to rebuild the topology.

> With the polygon boundaries displayed in the drawing, redisplay the node errors by executing the Noderror command on the Topology menu. To correct the error in the northwest polygon, zoom in to the error and use the AutoCAD Extend command to extend the horizontal line to the vertical line as shown in the following figure.

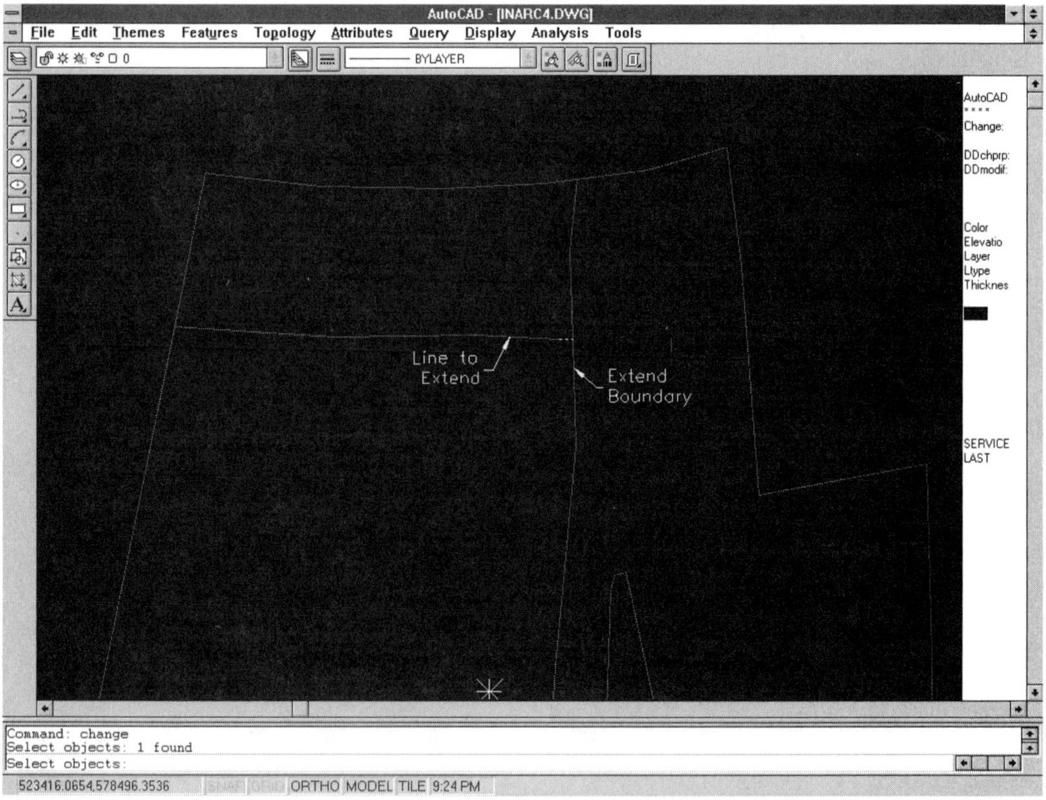

Correcting a node error.

With the feature corrected in AutoCAD, this new information must be written to the coverage and the topology rebuilt. To accomplish this, execute the Modfeat command found on the Features menu. The Modfeat command is used to update existing ArcCAD features. The Savefeat command can be used both to add new features to the data set as well as update existing features. The Delfeat command can be used to remove features from the data set. The command syntax for Modfeat is shown below.

```
Command: MODFEAT
Theme name (?/<LANDUSEL>): ENTER
```

Select the modified line.

```
Select objects: 1 found
Select objects: ENTER
Optional property table (?/<none>): ENTER
1 modified in C:\INARC\LANDUSE, 0 ignored.
```

Select Clean Now from the dialog box.

```
Output theme (?/<LANDUSEL>): ENTER
Dangle length <default: ENTER
Fuzzy tolerance <compute>: ENTER
Cleaning LANDUSEL.
Sorting...
Partial process enabled.  83 percent of the coverage
  will be processed.
Intersecting...
Assembling polygons...
  Sorting input file...
  Sorting label file...
  Processing...
  Assigning final IDs...
  Writing arc file...
  Generating polygon report...
Creating Polygon attribute table.
Dropping entity-feature links...
Unlinking...
```

By choosing Clean Now from the dialog box you have told ArcCAD to rebuild the topology for the coverage. You will notice from the previous output that the Clean command has rebuilt polygon topology.

To confirm the effects of the Modfeat command, erase the old features from the drawing and redisplay them from the theme using the Polys command found on the Display pulldown menu under Polygons.

```
Command: POLYS
Theme name (?/<LANDUSEL>): LANDUSE
Line symbol number (0-100) (?/): 5
```

To confirm that all dangle errors have been corrected, execute the Noderror command from the Topology menu.

```
Theme name (?/<LANDUSE>): LANDUSEL
Output report to <SCREEN>: ENTER
Error type? (Dangle/Pseudo/<Both>): ENTER
```

If no red squares appear on the screen, all dangle errors have been corrected.

Correcting Label Point Errors

The process of correcting all node errors itself has caused a new error. By correcting the error on the northwest polygon, you have created a new polygon. This new polygon has no label point.

To identify this, execute the Clrerror command found on the Topology menu to remove the node error symbols from the screen.

Next, execute the Laberror command from the Topology menu. You should see two stars now: one in the southeast polygon and a new one in the northwest polygon that was created when the node error was corrected.

LANDUSE polygons with label point errors.

Label point errors are simple to correct in ArcCAD. The Createlab command found on the Topology menu randomly assigns a unique ID to each polygon that does not have one. The only input required is the theme name and the ID base value. The ID base value sets the starting number for the User-IDs. This base value should be at least one higher than the highest User-ID currently in the theme or enter 0 to renumber all existing User-IDs.

The input for the Createlab command is shown below.

```
Command: CREATELAB
Theme name (?/<LANDUSE>): ENTER
User-ID base : 9
```

When the Createlab command is finished, the theme must be rebuilt to update the topology.

```
Command: BUILD
Input theme (?/<LANDUSE>): ENTER
Output theme (?/<LANDUSE>): ENTER
```

Select OK from the drop links dialog box.

```
Building polygons...
Processing...
  Sorting input file...
  Sorting label file...
  Processing...
  Assigning final IDs...
  Writing arc file...
  Generating polygon report...
Creating Polygon attribute table.
Dropping entity-feature links...
```

To view the theme with the new IDs, execute the Wbrowse command from the Attributes menu. The feature attribute table should look like the one in the following figure.

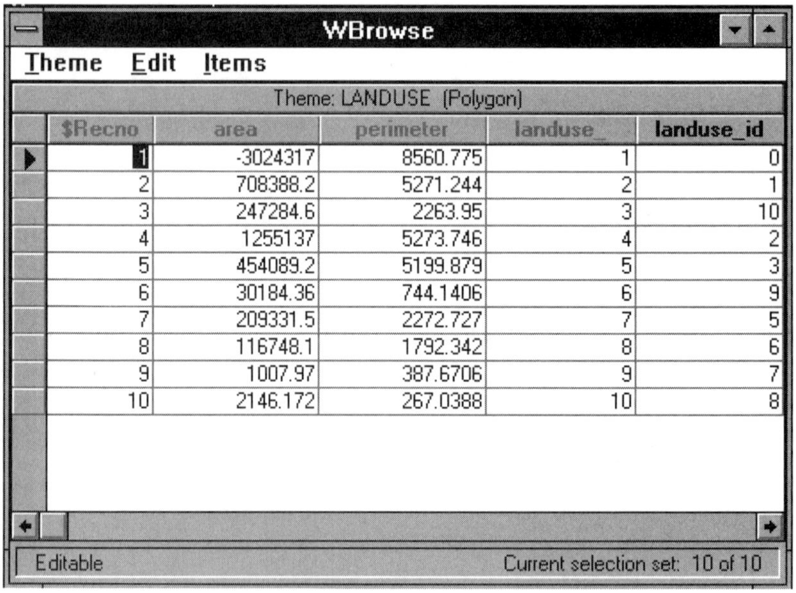

$Recno	area	perimeter	landuse_	landuse_id
1	-3024317	8560.775	1	0
2	708388.2	5271.244	2	1
3	247284.6	2263.95	3	10
4	1255137	5273.746	4	2
5	454089.2	5199.879	5	3
6	30184.36	744.1406	6	9
7	209331.5	2272.727	7	5
8	116748.1	1792.342	8	6
9	1007.97	387.6706	9	7
10	2146.172	267.0388	10	8

The LANDUSE FAT with new IDs.

To confirm the correction of the label errors, execute the Clrerror command and then reexecute the Laberror command. If no blue stars are displayed, all label errors have been corrected.

The Createlab command randomly assigns User-IDs to polygons. These IDs can be changed to user-specified numbers with two different commands for modifying the ID in the feature attribute table. The Idedit command then completes the update of the ID in the spatial database and feature entity data.

To update the ID assigned to the northwest polygon, execute the Modone command found on the Attributes menu under Records. At the theme name prompt, enter `LANDUSE`. At the select polygon prompt, pick a point in the northwest polygon.

The Modone command now displays the Attributes of the selected polygon in a dialog box.

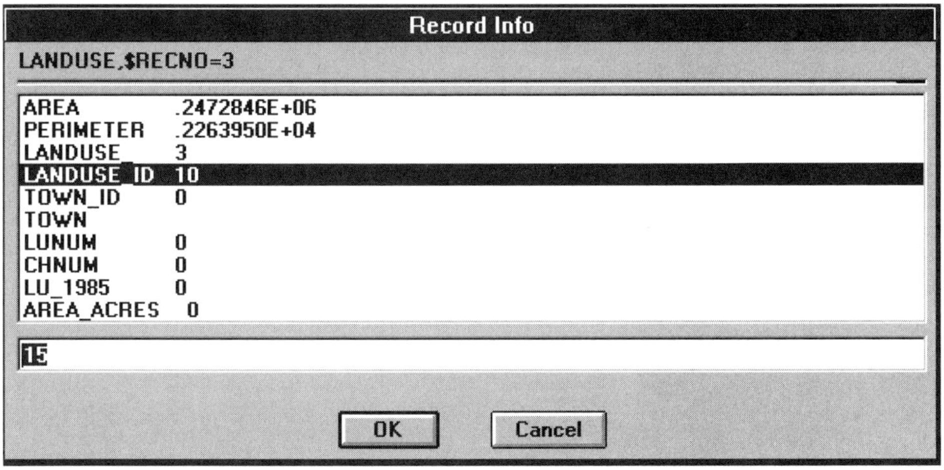

Modone dialog box for the northwest polygon.

To edit the User-ID field, single click on the LANDUSE_ID item. In the edit box type the number 15 and press Enter.

NOTE: *Be sure to press Enter when modifying item values within any of ArcCAD's dialog boxes. If you fail to press Enter and select OK instead, the new item value will not be written to the record.*

You have edited the User-ID in the LANDUSE themes feature attribute table, but the ID editing process is not complete. ArcCAD maintains User-IDs for all features in three separate locations. The first of the locations is the feature attribute table. The second location is the spatial database, and the third is as extended entity data.

To update the spatial database and the extended entity data with the new ID value, use the Idedit command. This command updates the spatial database and the extended entity data with the value in the feature attribute table.

You have completed the process of editing a polygon coverage. The command structure used is identical to that of editing line coverages, with the exception of having to define a new theme and create label points. The process of adding new polygons or lines to a coverage is identical, except you substitute Savefeat for Modfeat. The Savefeat command adds new lines to line and polygon coverages. Modfeat only changes lines that already exist within the coverage.

Modifying Lines

Now that you know the process to modify line coverages, let's correct the errors in the Streets coverage.

To begin, erase all entities on screen using the AutoCAD Erase command.

```
Command: ERASE
Select objects: 521132.0020,576349.8677
Other corner: 524082.2318,578892.7667
Select objects: ENTER
```

Next, create a new layer named CL and set it current.

```
Command: LA
LAYER
  ?/Make/Set/New/ON/OFF/Color/Ltype/Freeze/Thaw/LOck/
  Unlock: M
New current layer <0>: CL
?/Make/Set/New/ON/OFF/Color/Ltype/Freeze/Thaw/LOck/
  Unlock: ENTER
```

You now need to define a theme to access the street coverage. Use the Defthm command found on the Themes menu and follow the input shown below.

Defthm dialog box for Streets.

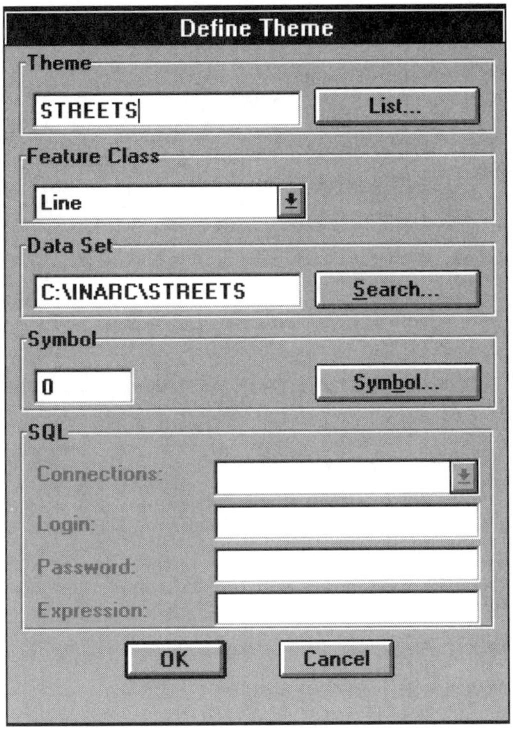

Now that you have prepared the work area, display the street centerlines in the drawing using the Arclines command.

```
Command: ARCLINES
Theme name (?/<STREETS>): ENTER
Display using? (Properties/Item/<Symbol>): ENTER
Line symbol number (0-100) (?/<0>): 1
```

As shown in the following figure, the lower portion of the map is missing two street centerline segments. In this section you will add these two segments to the theme. In the next chapter you will add the attribute data to these new street segments.

Centerline segments to be added.

To add the new lines to the theme, first draw the lines in using the AutoCAD Line command. Use the AutoCAD endpoint snap to select the endpoints of the existing centerlines. With the new lines drawn, set the User-ID for the new features to be added with the User-ID command and use the Addfeat command to add the features to the data set. Both of these commands can be found on the features menu.

```
Command: USERID
Theme name (?/<STREETS>): ENTER
User ID base <1>:50
User ID increment <1>: ENTER
Command: SAVEFEAT
Theme name (?/<STREETS>): ENTER
Select objects: Selected the two new lines
```

```
Optional property table (<?/<none>): ENTER
2 written to C:\INARC\STREETS, 0 modified, 0 ignored.
Output theme (?/<STREETS>): ENTER
```

Select Clean Now from the dialog box.

```
Dangle length <default>: 10
Fuzzy tolerance <compute>: 10
Cleaning STREETS.
Sorting...
Partial process enabled.  86 percent of the coverage
  will be processed.
Intersecting...
Assembling polygons...
  Sorting input file...
  Sorting label file...
  Processing...
  Assigning final IDs...
  Writing arc file...
  Generating polygon report...
Creating Polygon attribute table.
Re-building arc attribute table...
Creating Line attribute table.
Dropping entity-feature links...
Unlinking...
```

With the Addfeat command complete, the two new lines have been added to the theme. To confirm this, Erase all entities from the screen and redisplay the centerlines using the Arclines command found on the Display menu under Lines.

```
Command: ARCLINES
Theme name (?/<STREETS>): ENTER
Display using? (Properties/Item/<Symbol>):
Line symbol number (0-100) (?/<0>): 1
```

The completed drawing is shown in the following figure.

The Street centerlines with new segments.

Review

You have learned how to modify spatial features within themes. You should be familiar with how to access both line and polygon information in a polygon coverage, as well as how to modify, add, and delete features from a coverage. You have also learned how to identify errors in line and polygon coverages and how to update and modify feature IDs.

Chapter

Linking Tabular Data

Adding Descriptions to Your Graphics

This chapter covers manipulating items within and adding data to feature attribute tables (FATs). Throughout this section you will work exclusively with the Attributes menu. Through this menu you have access to all of the ArcCAD database manipulation functions. You will learn the following database manipulations:

- ❒ Add items to a FAT
- ❒ Delete items from a FAT
- ❒ Calculate new items
- ❒ Move items from one table to another
- ❒ Add values to database items.

Setup

To prepare for this chapter, follow these steps.

Start AutoCAD and open the drawing named INARC6.DWG.

Load ArcCAD if it is not set to load automatically. The drawing should look like the following figure.

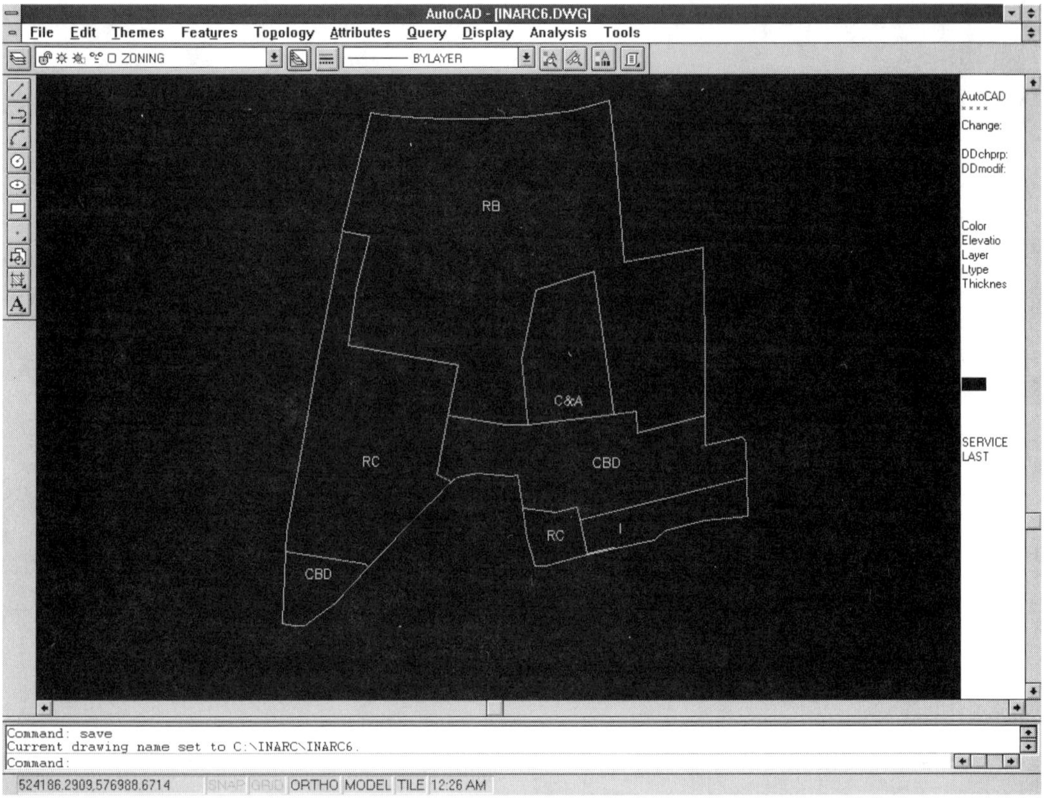

The INARC6 drawing file.

The INARC6 drawing contains a theme for the zone polygons, and the coverage has been created for you.

Databases, Items, and Records

In order to effectively work with GIS in general and ArcCAD specifically, you need to have a clear understanding of how a database works. Within a database there are two typical components: items, representing the columns in a table; and records, representing the rows in a table.

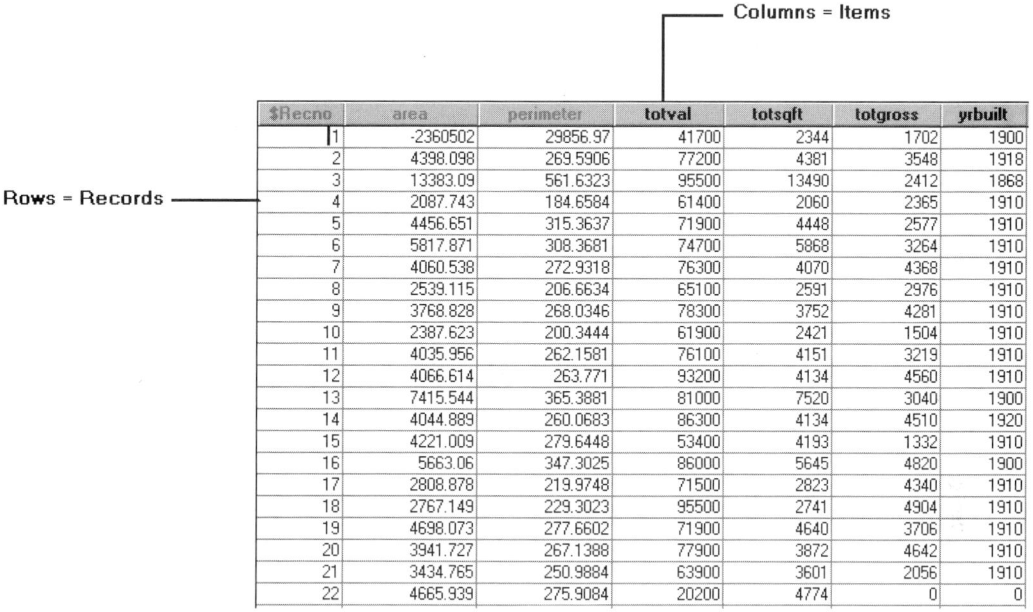

$Recno	area	perimeter	totval	totsqft	totgross	yrbuilt
1	-2360502	29856.97	41700	2344	1702	1900
2	4398.098	269.5906	77200	4381	3548	1918
3	13383.09	561.6323	95500	13490	2412	1868
4	2087.743	184.6584	61400	2060	2365	1910
5	4456.651	315.3637	71900	4448	2577	1910
6	5817.871	308.3681	74700	5868	3264	1910
7	4060.538	272.9318	76300	4070	4368	1910
8	2539.115	206.6634	65100	2591	2976	1910
9	3768.828	268.0346	78300	3752	4281	1910
10	2387.623	200.3444	61900	2421	1504	1910
11	4035.956	262.1581	76100	4151	3219	1910
12	4066.614	263.771	93200	4134	4560	1910
13	7415.544	365.3881	81000	7520	3040	1900
14	4044.889	260.0683	86300	4134	4510	1920
15	4221.009	279.6448	53400	4193	1332	1910
16	5663.06	347.3025	86000	5645	4820	1900
17	2808.878	219.9748	71500	2823	4340	1910
18	2767.149	229.3023	95500	2741	4904	1910
19	4698.073	277.6602	71900	4640	3706	1910
20	3941.727	267.1388	77900	3872	4642	1910
21	3434.765	250.9884	63900	3601	2056	1910
22	4665.939	275.9084	20200	4774	0	0

A typical database.

An *item* in a database is a column heading. Each item in the database contains data values that describe a feature. A group of item values describing a single feature creates a *record.*

In ArcCAD, every spatial feature has a single record describing it in the FAT. When a new feature is created, a record is added to the FAT. If a feature is deleted from the spatial database, a record is removed from the FAT. When a new record is added to the FAT, ArcCAD calculates the default items for the feature. The default items created are dependent on the feature class of the information being added. The following table defines the default items for each feature class.

Polygon	Area	Perimeter	Cov_	Cov_ID		
Point	Area	Perimeter	Cov_	Cov_ID		
Line	Fnode_	Tnode_	Lpoly_	Rpoly_Length	Cov_	Cov_ID

When adding items to a FAT, the order and position of these default items should never be modified, that is, no user-defined item should ever be placed in between or before these items.

When manipulating records within a FAT, the default items values should never be changed, with the possible exception of the User-IDs. Each of the default item's values are calculated values and changing the values will have an adverse effect on the theme.

Manipulating Items in a Feature Attribute Table

In this section you will learn to add, edit, and delete items in a FAT. The commands and methods used in this section are identical to the editing of any database in ArcCAD, including lookup tables and property tables, which are covered in the next chapter.

Adding an Item

You will begin this editing session by adding an item to the ZONING theme. The item name will be ZONE. Once created, item values will be added to the theme. This section will demonstrate several methods for coordinating spatial data with attribute data.

The first step is to add the item ZONE to the ZONING FAT. This is done with the Additem command found on the Attributes pulldown menu under Items.

```
Command: ADDITEM
Theme name (?/<THEME>): ZONING
Item name (?): ZONE
Item width: 6
Item type?  (Character/Date/<Numeric>): C
Start item <last>: ENTER
```

The command has added an item named ZONE to the FAT. The item value can be up to six characters long and alphanumeric. Accepting the default value of LAST at the Start item prompt told ArcCAD to place the ZONE item after the last item in the FAT.

With the item added to the table, you can now add values to the item. Each ZONE polygon will receive a value.

Now you need to coordinate the spatial information with its corresponding attribute information. ArcCAD provides two methods of accomplishing this. The first is through the use of the Modone command. Although straightforward and effective, the Modone command is slow. The other method is to write the User-ID of each polygon to the drawing. In effect, this gives you a link item to coordinate with. Let's look at both methods now.

> To begin, execute the Modone command from the Attributes pulldown menu under Records. At the theme name prompt accept the default of ZONING.
>
> When prompted to select a polygon, click in the polygon labeled RB. ArcCAD presents the database record for this polygon in a dialog box as shown below.

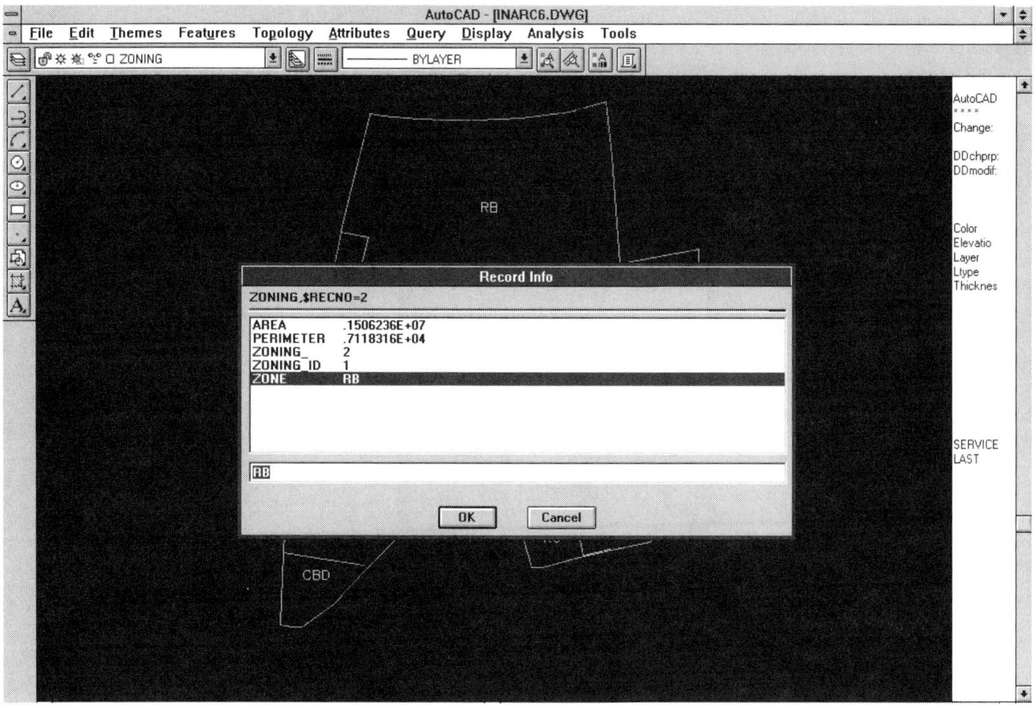

The Modone dialog box for a ZONING polygon.

Single click on the ZONE item. Type RB into the edit box at the bottom of the dialog box and press Enter. The item value will be updated, and you may select OK to finish the command.

Congratulations, you have just added your first item to a FAT and placed a value in it. As you can see, if you had several hundred polygons to add values to, this could be a time-consuming method. Thankfully, ArcCAD provides a faster method.

The faster method of adding data to a theme is to display the User-IDs into the drawing to use as a link item. The User-ID can then be used to identify the correct record to which to add the data.

To begin this method, execute the Polytxt command found on the DISPLAY menu under Polygons.

```
Command: LABELTXT
Theme name (?/<ZONING>): ENTER
Item (?): ?
```

Select ZONING_ID from the Items Selection dialog box.

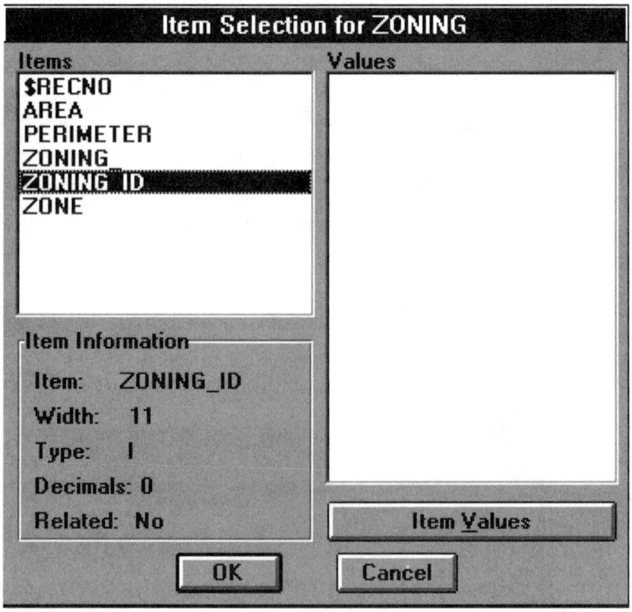

```
Optional lookup table (?/<none>): ENTER
Alignment?
  (Center/Middle/Right/TL/TC/TR/ML/MC/MR/BL/BC/BR/
  <Left>): ENTER
Text symbol number (0-100) (?/<0>): ENTER
```

The drawing should now look like the following figure.

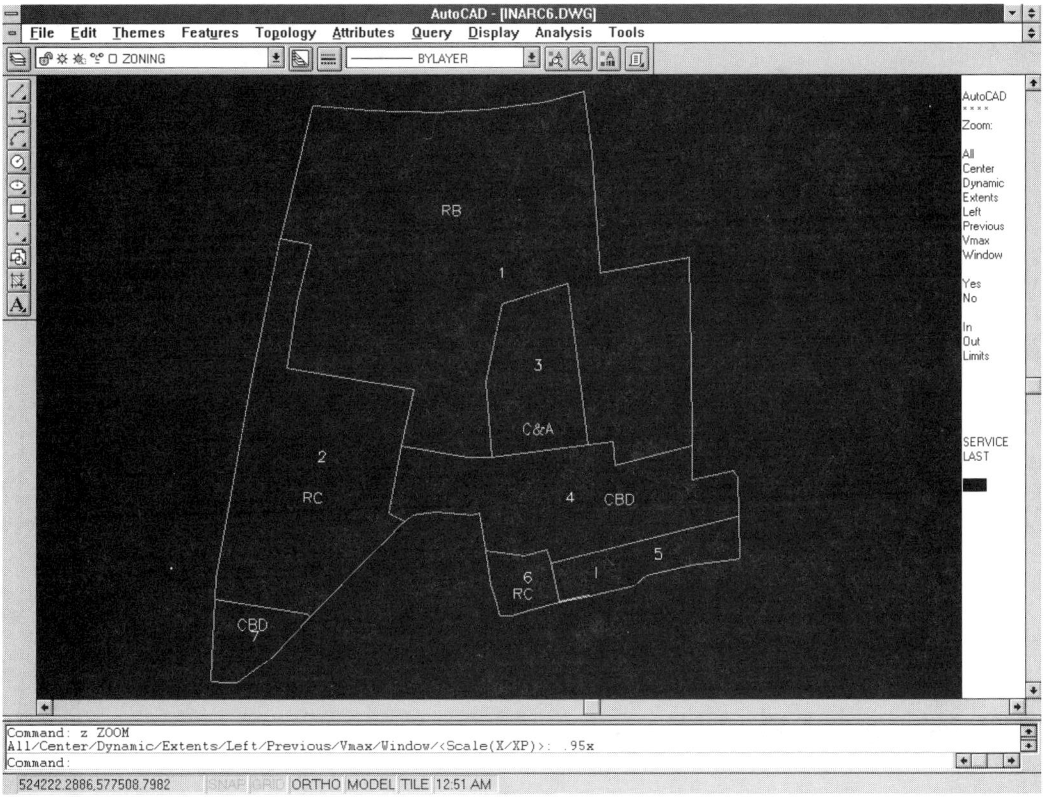

INARC6 drawing file with User-IDs.

The User-IDs are displayed in the drawing, identifying each ZONE's User-ID. Now, the Wbrowse command can be used to add values for the ZONE item.

Execute the Wbrowse command and assign the following values.

User-ID	Zone
2	RC
3	C&A
4	CBD
5	I
6	RC
7	CBD

Once all of the values have been added, select THEMES from the Wbrowse dialog box and choose Save and Exit.

As you can see, the second method of adding item values is considerably faster than using the Modone command. For larger themes it is recommended that a plot of the drawing be created. From this plot the values can be added.

NOTE: *The Wbrowse command is new to ArcCAD 11.4. For 11.3 or earlier versions, the Modrec command found on the Attributes pulldown menu under Records must be used to add item values to FATs.*

The process of adding item values to point and line themes is identical to the process described previously.

Editing and Deleting Items

Now that you have learned the basics of adding information to FATs, you will look at the commands available for manipulating these tables. The first thing covered in this section is how to reduce the size of FATs to make them more manageable.

These commands will allow you to pull items from the FAT and place them in a record theme. In later chapters you will learn how to temporarily link these record themes back to the FAT for query and analysis, as well as how to merge the record theme back into the FAT.

You will also use commands that allow you to calculate item values, move items within a table, and change item definitions and delete items.

Using Record Themes

The PARCELS theme currently contains 12 items. Although all of these items contain useful information, large numbers of items can become unmanageable. To solve this problem you can pull some of the items from the FAT and place them in a record theme. The record theme can then be related to the FAT when they are needed. In this section you will pull five of the items from the FAT and place them in a record theme named LOTINFO.

Execute the Pullitem command found on the Attributes pulldown menu under Items. The syntax and items to pull are shown below.

```
Command: PULLITEM
Input theme (?/<PARCELS>): ENTER
Output theme (?): LOTINFO
Theme LOTINFO is not defined.
```

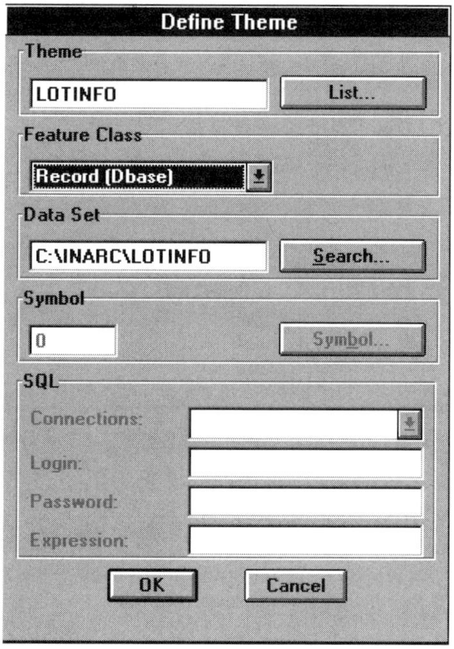

Defthm dialog box for LOTINFO.

Select OK from the Define Theme dialog box.

```
Defining LOTINFO...
Items to pull (?/<END>): PARCELS_ID
Items to pull (?/<END>): OWNER1
Items to pull (?/<END>): ADDRESS
Items to pull (?/<END>): OCCUP
Items to pull (?/<END>): YRBUILT
Items to pull (?/<END>): ENTER
```

The PARCELS_ID field was pulled from the FAT to provide a common item to link when relating the tables together later. To view the data in the LOTINFO theme, execute the Wbrowse command and specify LOTINFO at the theme name prompt.

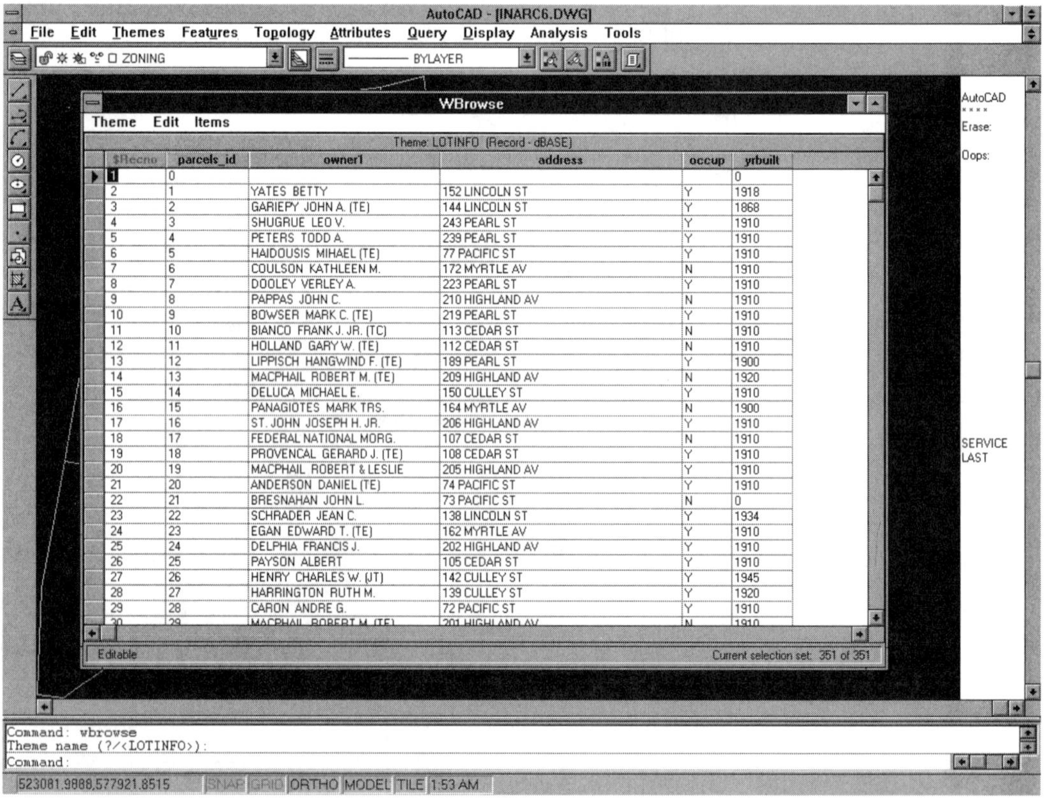

The LOTINFO record theme.

The Pullitem command made a copy of the items and their values in the LOTINFO record theme. To finish the process, these items must now be dropped from the FAT. This is done by using the Dropitem command found on the Attributes pulldown menu under Items. The input for the Dropitem command is shown below.

```
Command: DROPITEM
Theme name (?/<parcels>): ENTER
Item name (?): ?
```

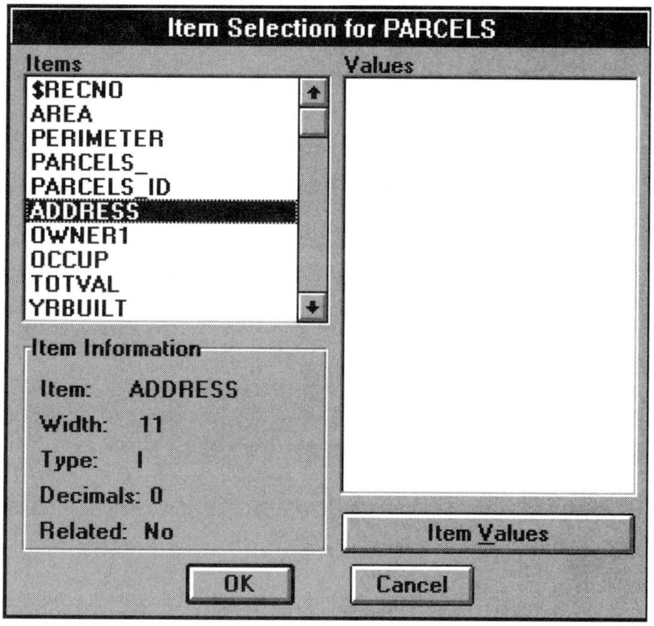

Select Address from the Item Selection dialog box.

```
ADDRESS
Dropping ITEM ADDRESS...
```

Repeat the Dropitem command for the other three items. Be sure not to drop the PARCELS_ID field from the FAT.

The PARCELS FAT should now look like the following figure.

WBrowse

Theme Edit Items

Theme: PARCELS (Polygon)

$Recno	area	perimeter	parcels_	parcels_id	totval	totsqft	totgross	image
1	-.2360502E+07	.2985697E+05	1	309	41700	2344	1702	
2	.4398098E+04	.2695906E+03	2	1	77200	4381	3548	
3	.1338309E+05	.5616323E+03	3	2	95500	13490	2412	
4	.2087743E+04	.1846584E+03	4	3	61400	2060	2365	
5	.4456651E+04	.3153637E+03	5	4	71900	4448	2577	
6	.5817871E+04	.3083681E+03	6	5	74700	5868	3264	
7	.4060538E+04	.2729318E+03	7	6	76300	4070	4368	
8	.2539115E+04	.2066634E+03	8	7	65100	2591	2976	
9	.3768828E+04	.2680346E+03	9	8	78300	3752	4281	
10	.2387623E+04	.2003444E+03	10	9	61900	2421	1504	
11	.4035956E+04	.2621581E+03	11	10	76100	4151	3219	
12	.4066614E+04	.2637710E+03	12	11	93200	4134	4560	
13	.7415544E+04	.3653881E+03	13	12	81000	7520	3040	
14	.4044889E+04	.2600683E+03	14	13	86300	4134	4510	
15	.4221009E+04	.2796448E+03	15	14	53400	4193	1332	
16	.5663060E+04	.3473025E+03	16	15	86000	5645	4820	
17	.2808878E+04	.2199748E+03	17	16	71500	2823	4340	
18	.2767149E+04	.2293023E+03	18	17	95500	2741	4904	
19	.4698073E+04	.2776602E+03	19	18	71900	4640	3706	
20	.3941727E+04	.2671388E+03	20	19	77900	3872	4642	
21	.3434765E+04	.2509884E+03	21	20	63900	3601	2056	
22	.4665939E+04	.2759084E+03	22	21	20200	4774	0	
23	.4359919E+04	.3191035E+03	23	22	66900	4480	2062	
24	.6359672E+04	.3217514E+03	24	23	72900	6428	3651	
25	.3399705E+04	.2690635E+03	25	24	70600	3401	4290	
26	.3117594E+04	.2412397E+03	26	25	54300	3057	1944	

Editable Current selection set: 351 of 351

PARCELS FAT after using Dropitem command.

Calculating an Item Value

The last database operation you will do is calculate the square foot value of the parcels and place this value in a new item. You must first create a target item in the table to hold the per square foot value. Once the item is created, the value can be calculated. The input for this is shown below.

```
Command: additem
Theme name (?/<parcels>): ENTER
Item name (?): sqftval
Item width: 10
Item type?  (Character/Date/<Numeric>):
Item decimal places <0>: 2
Start item <last>: ENTER
Command: CALCULATE
Theme name (?/<parcels>): ENTER
Target Item (?): ?
```

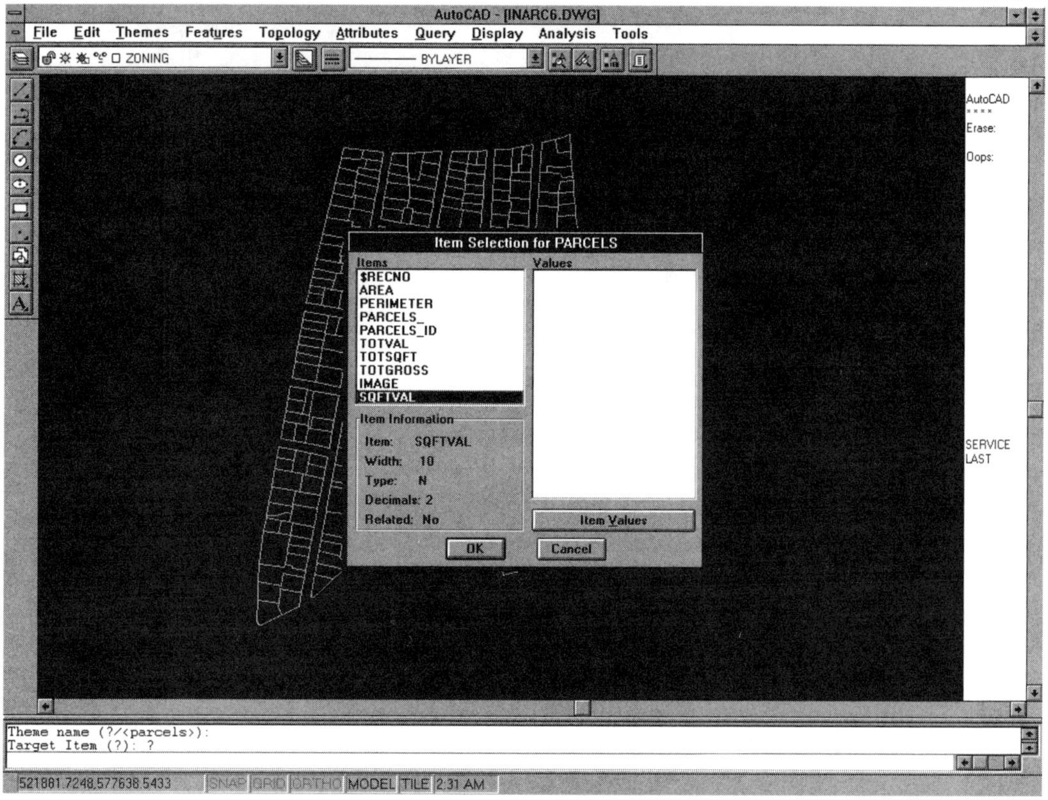

The Item Selection dialog box.

Select the target item, SQFTVAL, from the Item Selection dialog box.

Once the target item is selected, ArcCAD provides an arithmetic expression builder to specify the calculations.

Equation for SQFTVAL.

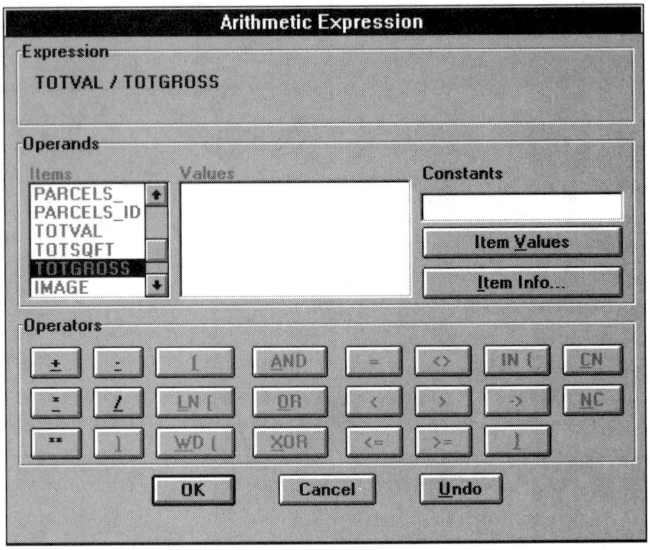

Using the expression builder, create a calculation to divide TOTVAL by TOTGROSS. The result of this expression will be placed on the SQFTVAL item.

Once the calculation is entered as shown above, select OK from the dialog box.

To view the new data, use Wbrowse from the Atributes Datafiles menu.

The PARCELS FAT is shown as follows.

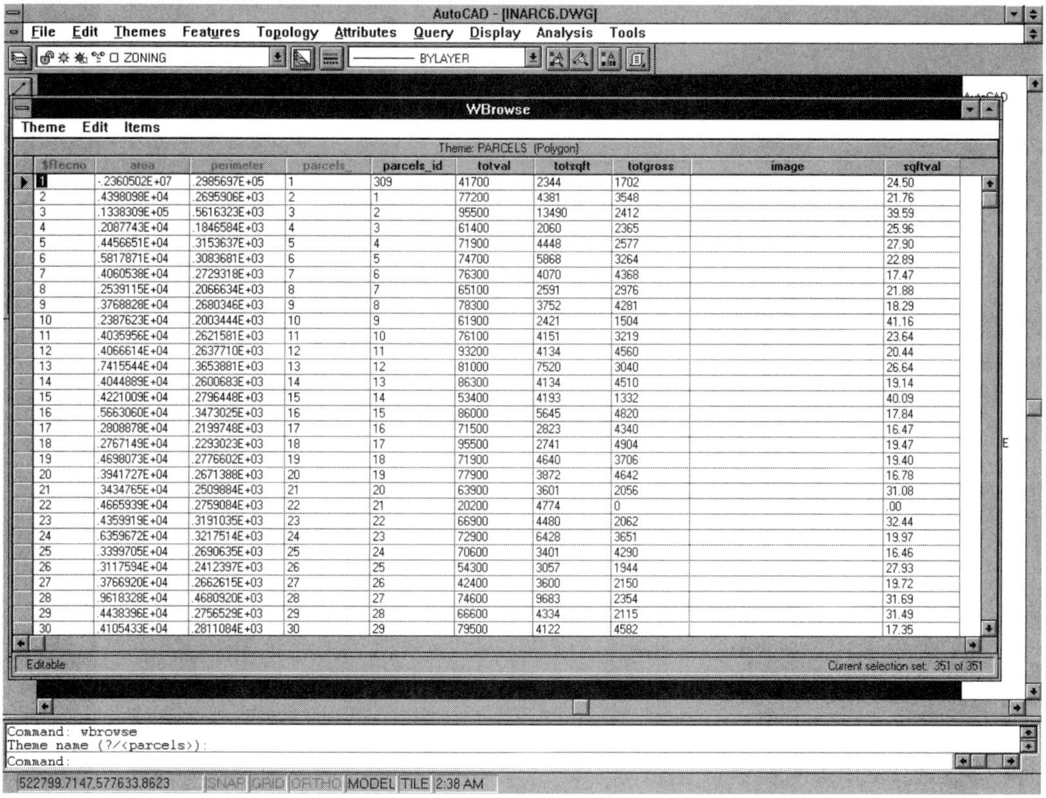

$Recno	area	perimeter	parcels_	parcels_id	totval	totsqft	totgross	image	sqftval
1	-.2360502E+07	.2985697E+05	1	309	41700	2344	1702		24.50
2	.4398098E+04	.2695906E+03	2	1	77200	4381	3548		21.76
3	.1338309E+05	.5616323E+03	3	2	95500	13490	2412		39.59
4	.2087743E+04	.1846584E+03	4	3	61400	2060	2365		25.96
5	.4456651E+04	.3153637E+03	5	4	71900	4448	2577		27.90
6	.5817871E+04	.3083681E+03	6	5	74700	5868	3264		22.89
7	.4060538E+04	.2729318E+03	7	6	76300	4070	4368		17.47
8	.2539115E+04	.2066634E+03	8	7	65100	2591	2976		21.88
9	.3768828E+04	.2680346E+03	9	8	78300	3752	4281		18.29
10	.2387623E+04	.2003444E+03	10	9	61900	2421	1504		41.16
11	.4035956E+04	.2621581E+03	11	10	76100	4151	3219		23.64
12	.4066614E+04	.2637710E+03	12	11	93200	4134	4560		20.44
13	.7415544E+04	.3653881E+03	13	12	81000	7520	3040		26.64
14	.4044889E+04	.2600683E+03	14	13	86300	4134	4510		19.14
15	.4221009E+04	.2796448E+03	15	14	53400	4193	1332		40.09
16	.5663060E+04	.3473025E+03	16	15	86000	5645	4820		17.84
17	.2808878E+04	.2199748E+03	17	16	71500	2823	4340		16.47
18	.2767149E+04	.2293023E+03	18	17	95500	2741	4904		19.47
19	.4698073E+04	.2776602E+03	19	18	71900	4640	3706		19.40
20	.3941727E+04	.2671388E+03	20	19	77900	3872	4642		16.78
21	.3434765E+04	.2509884E+03	21	20	63900	3601	2056		31.08
22	.4665939E+04	.2759084E+03	22	21	20200	4774	0		.00
23	.4359919E+04	.3191035E+03	23	22	66900	4480	2062		32.44
24	.6359672E+04	.3217514E+03	24	23	72900	6428	3651		19.97
25	.3399705E+04	.2690635E+03	25	24	70600	3401	4290		16.46
26	.3117594E+04	.2412397E+03	26	25	54300	3057	1944		27.93
27	.3766920E+04	.2662615E+03	27	26	42400	3600	2150		19.72
28	.9618328E+04	.4680920E+03	28	27	74600	9683	2354		31.69
29	.4438396E+04	.2756529E+03	29	28	66600	4334	2115		31.49
30	.4105433E+04	.2811084E+03	30	29	79500	4122	4582		17.35

PARCELS FAT after Calculate.

Review

You have completed the exercise on manipulating FATs. In this chapter you learned how to add and delete items to a FAT, calculate new items, move items from one table to another, and add values to database items. In the next chapter you will learn to create new record themes and use the themes as relate tables and lookup tables.

Chapter 7

Working with Tables

Lookup and Relate Tables

You should now be familiar with feature attribute tables (FATs). These tables consist of database records with direct links to spatial information. In this chapter you will be introduced to several other kinds of tables that are integral to a GIS. You will learn how to create and manipulate dBASE III files for use as lookup tables and relate tables, how to directly access non-dBASE format files using SQL, and how to convert other file formats into dBASE files. In addition to creating access to these files, you will also learn how to temporarily and permanently join these tables with FATs.

Setup

To prepare for this chapter,

> Start AutoCAD and open the INARC7 drawing file.
>
> Load ArcCAD if it is not set to load automatically. The drawing has two themes defined in it for the UTILS coverage and the PARCELS coverage.

Record Themes

Record themes are ArcCAD's link to nongraphic attribute tables. Record themes are links to database files that can be created, read, and modified from within ArcCAD. These database files have no links or relationships with AutoCAD entities and may or may not have any relationship with other ArcCAD tables.

When defining a record theme, ArcCAD will search for an existing database table with the path and name specified. If existing data is not found, ArcCAD will create a new table. If data is found, ArcCAD will connect to it and allow the table to be viewed, queried, and modified.

Previously you created two separate nongraphic based tables, the LOTINFO table created in the last chapter and the UTILSOPT table created in Chapter 5.

Let's begin by defining themes for these two tables.

From the Themes menu, choose Defthm and fill out the dialog box as shown in the following figure.

Defthm for LOTINFO database table.

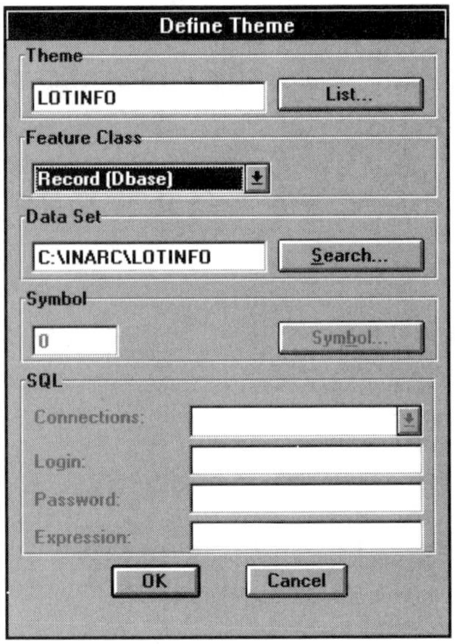

Next, define a theme for the UTILSOPT table.

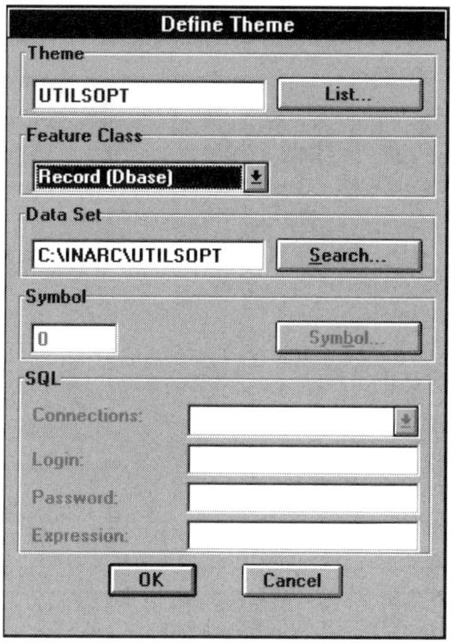

Defthm for UTILSOPT table.

Use the Wbrowse command on the Attributes menu to view the two tables, which should look like the following figures.

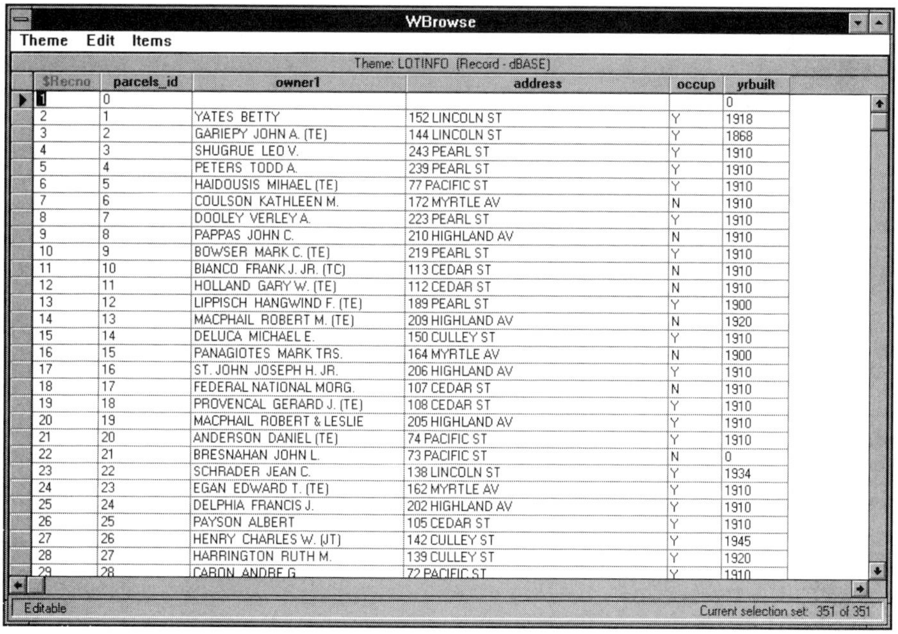

$Recno	parcels_id	owner1	address	occup	yrbuilt
1	0				0
2	1	YATES BETTY	152 LINCOLN ST	Y	1918
3	2	GARIEPY JOHN A. (TE)	144 LINCOLN ST	Y	1868
4	3	SHUGRUE LEO V.	243 PEARL ST	Y	1910
5	4	PETERS TODD A.	239 PEARL ST	Y	1910
6	5	HAIDOUSIS MIHAEL (TE)	77 PACIFIC ST	Y	1910
7	6	COULSON KATHLEEN M.	172 MYRTLE AV	N	1910
8	7	DOOLEY VERLEY A.	223 PEARL ST	Y	1910
9	8	PAPPAS JOHN C.	210 HIGHLAND AV	N	1910
10	9	BOWSER MARK C. (TE)	219 PEARL ST	Y	1910
11	10	BIANCO FRANK J. JR. (TC)	113 CEDAR ST	N	1910
12	11	HOLLAND GARY W. (TE)	112 CEDAR ST	N	1910
13	12	LIPPISCH HANGWIND F. (TE)	189 PEARL ST	Y	1900
14	13	MACPHAIL ROBERT M. (TE)	209 HIGHLAND AV	N	1920
15	14	DELUCA MICHAEL E.	150 CULLEY ST	Y	1910
16	15	PANAGIOTES MARK TRS.	164 MYRTLE AV	N	1900
17	16	ST. JOHN JOSEPH H. JR.	206 HIGHLAND AV	Y	1910
18	17	FEDERAL NATIONAL MORG.	107 CEDAR ST	N	1910
19	18	PROVENCAL GERARD J. (TE)	108 CEDAR ST	Y	1910
20	19	MACPHAIL ROBERT & LESLIE	205 HIGHLAND AV	Y	1910
21	20	ANDERSON DANIEL (TE)	74 PACIFIC ST	Y	1910
22	21	BRESNAHAN JOHN L.	73 PACIFIC ST	N	0
23	22	SCHRADER JEAN C.	138 LINCOLN ST	Y	1934
24	23	EGAN EDWARD T. (TE)	162 MYRTLE AV	Y	1910
25	24	DELPHIA FRANCIS J.	202 HIGHLAND AV	Y	1910
26	25	PAYSON ALBERT	105 CEDAR ST	Y	1910
27	26	HENRY CHARLES W. (JT)	142 CULLEY ST	Y	1945
28	27	HARRINGTON RUTH M.	139 CULLEY ST	Y	1920
29	28	CARON ANDRE G.	72 PACIFIC ST	Y	1910

The LOTINFO table.

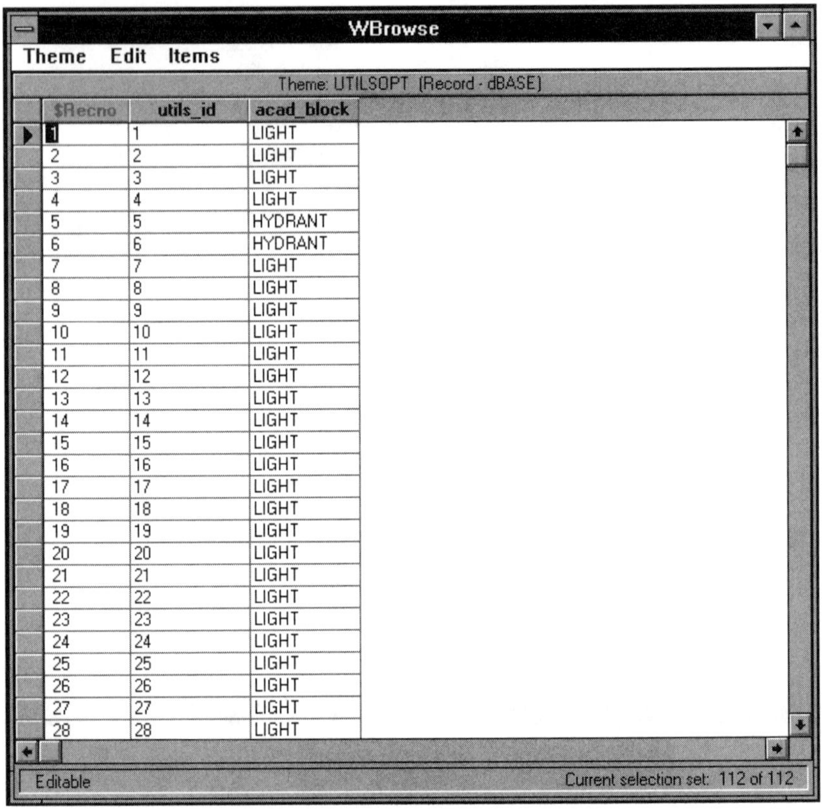

$Recno	utils_id	acad_block
1	1	LIGHT
2	2	LIGHT
3	3	LIGHT
4	4	LIGHT
5	5	HYDRANT
6	6	HYDRANT
7	7	LIGHT
8	8	LIGHT
9	9	LIGHT
10	10	LIGHT
11	11	LIGHT
12	12	LIGHT
13	13	LIGHT
14	14	LIGHT
15	15	LIGHT
16	16	LIGHT
17	17	LIGHT
18	18	LIGHT
19	19	LIGHT
20	20	LIGHT
21	21	LIGHT
22	22	LIGHT
23	23	LIGHT
24	24	LIGHT
25	25	LIGHT
26	26	LIGHT
27	27	LIGHT
28	28	LIGHT

The UTILSOPT table.

The main uses for record themes within ArcCAD are as optional property tables, lookup tables, and relate or join tables. Once a property table has been created, it can later be permanently merged with or temporarily related to the FAT.

Merging Tables

ArcCAD provides the capability to merge database tables based on common items or an item value and a record number. Any tables accessible by ArcCAD can be related. For example, a record theme representing a property table can be merged with a FAT, or two record themes with a common item can be merged.

In the sample project data, the UTILSOPT table is a property table. It was created by extracting the block name property from the hydrants and street lights from the drawing in Chapter 5. Once the data has been extracted to the table and the coverage has been created and built, the property table can be merged with the FAT.

In this exercise you will merge the UTILSOPT table into the UTILS FAT. The command used to merge tables is Joinitem, found on the Attributes menu under Items. Follow the command input below to merge these two tables.

```
Command: JOINITEM
Input theme (?/<UTILSOPT>): UTILS
Join theme (?): UTILSOPT
Output theme (?/<UTILS>): ENTER
Relate item (?): ?
```

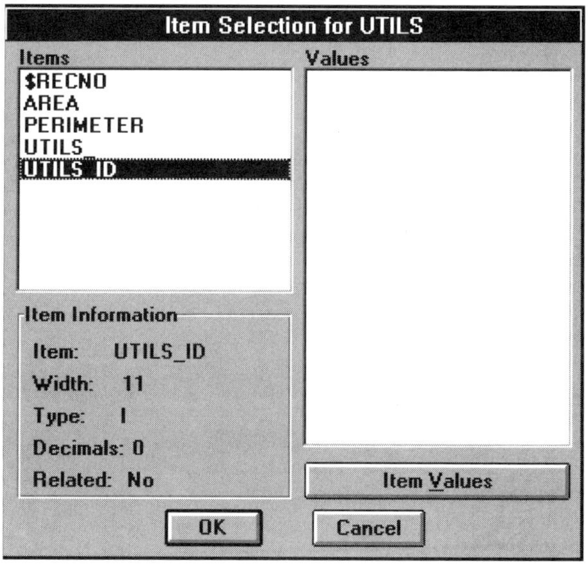

Select UTILS_ID from the Item Selection dialog box.

Select UTILS_ID from the Items list.

```
Start item (?/<last>): ENTER
Join method? (Link/Ordered/<LInear>): ENTER
Item "UTILS_ID" duplicated, Join File version dropped
```

The UTILSOPT table has now been merged into the UTILS FAT. The UTILS_ID item from the UTILSOPT table has been dropped because it already existed in the UTILS FAT. The UTILSOPT table has been left completely in tact. You may now delete the UTILSOPT table or save it for later use. The new UTILS FAT should look like the one shown in the following figure.

WBrowse

Theme Edit Items

Theme: UTILS (Point)

$Recno	area	perimeter	utils_	utils_id	acad_block
1	.0000000E+00	.0000000E+00	1	1	LIGHT
2	.0000000E+00	.0000000E+00	2	2	LIGHT
3	.0000000E+00	.0000000E+00	3	3	LIGHT
4	.0000000E+00	.0000000E+00	4	4	LIGHT
5	.0000000E+00	.0000000E+00	5	5	HYDRANT
6	.0000000E+00	.0000000E+00	6	6	HYDRANT
7	.0000000E+00	.0000000E+00	7	7	LIGHT
8	.0000000E+00	.0000000E+00	8	8	LIGHT
9	.0000000E+00	.0000000E+00	9	9	LIGHT
10	.0000000E+00	.0000000E+00	10	10	LIGHT
11	.0000000E+00	.0000000E+00	11	11	LIGHT
12	.0000000E+00	.0000000E+00	12	12	LIGHT
13	.0000000E+00	.0000000E+00	13	13	LIGHT
14	.0000000E+00	.0000000E+00	14	14	LIGHT
15	.0000000E+00	.0000000E+00	15	15	LIGHT
16	.0000000E+00	.0000000E+00	16	16	LIGHT
17	.0000000E+00	.0000000E+00	17	17	LIGHT
18	.0000000E+00	.0000000E+00	18	18	LIGHT
19	.0000000E+00	.0000000E+00	19	19	LIGHT
20	.0000000E+00	.0000000E+00	20	20	LIGHT
21	.0000000E+00	.0000000E+00	21	21	LIGHT
22	.0000000E+00	.0000000E+00	22	22	LIGHT
23	.0000000E+00	.0000000E+00	23	23	LIGHT

Editable Current selection set: 112 of 112

UTILS FAT after using Joinitem command.

Join and Relate Methods

The Joinitem process prompted for a join method. ArcCAD supports four methods for merging and relating tables. The methods are Link, Ordered, Linear, and Table.

Link: Only the *from* theme must contain the relate item. The *to* theme can be sorted in any order. The item value of the relate item in the *from* theme specifies the record number in the *to* theme that is to be merged.

Ordered: A binary search is used on the *to* theme to find matching records. This requires that the *to* theme be sorted on the relate item in ascending order. Both the *from* theme and the *to* theme must contain the relate item.

Linear: A linear search of the *to* theme is performed to find matching records. The relate item must exist in both themes, with the same item definitions. Both themes can be sorted in any order.

Table (Relate only): The *to* theme is a lookup table sorted in ascending order on the relate item. Both tables must contain the relate item. The records in the *from* theme are matched to records in the *to* theme, using a binary search. If no match is found, the record in the *from* theme is matched to the next highest value in the *to* theme. If there is no higher value in the *to* theme, the *from* theme record is matched to the last record in the *to* theme.

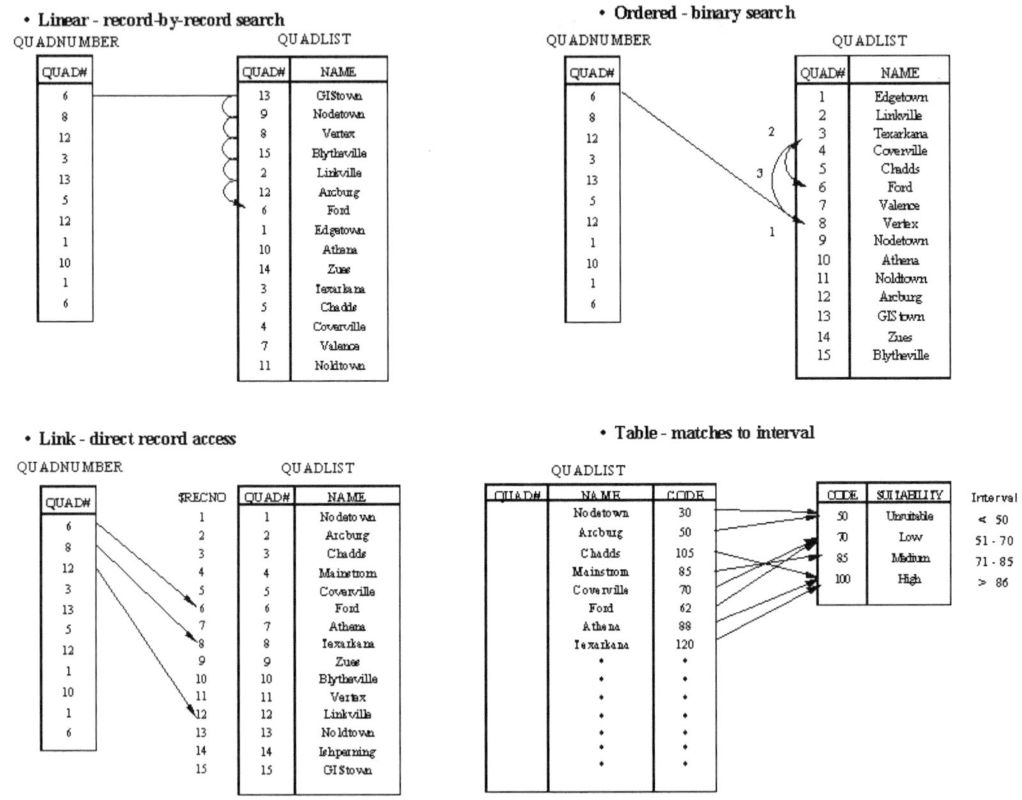

Join and Relate methods.

Linear is the default method. When doing a relate or a join, linear will be the slowest method if both themes are not sorted on the relate item. If both tables are sorted on the relate item, the linear method will be the fastest to process. The ordered method is generally the fastest method of relating or joining tables.

Relating Tables (Temporary Joins)

The process of relating tables in ArcCAD is similar to joining tables. The difference is that a relate is temporary. The relate exists only for the current drawing session and only one relate can exist at a time.

Once two tables have been related, the items from the relate *to* theme will be displayed in the relate *from* theme. The related items in the *from* theme will be displayed with a # preceding the item name. When querying based on a related item, it is important to include the # before the item name.

The sample project data includes a relate table for the PARCELS coverage called LOTINFO. LOTINFO was created in the previous chapter by taking items from the FAT. Pulling the items from the FAT is beneficial in several ways.

- ❐ The FAT remains smaller and therefore faster.
- ❐ Only the items that are needed for a particular task need to be shown in the table.
- ❐ Each user of the PARCELS coverage can maintain the attributes independent of the PARCELS theme itself.

As an example, in a typical municipal GIS, both the planning department and the assessors department may need access to parcel information. The planning office could maintain one database with a common item in the PARCELS FAT. The planning department can update and modify the external table using any database package capable of reading the DBF format. At the same time, the assessors department can be maintaining their own table. When one of the departments needs to do spatial analysis based on these attributes, they can simply relate their table to the FAT and do the analysis.

One more benefit of the relate capability is that only the user of a particular set of attributes has access to and responsibility for those attributes. For instance, in the previous example the assessors department would not have access to the planning department's data. This increases the level of database integrity in an enterprise-wide solution.

To relate the PARCELS FAT to the LOTINFO record theme, follow the input below.

```
Command: RELATE
Relate? (ON/<OFF>): ON
From theme (?/<UTILS>): PARCELS
To theme (?): LOTINFO
Relate item (?): ?
```

Select PARCELS_ID from the Item Selection dialog box.

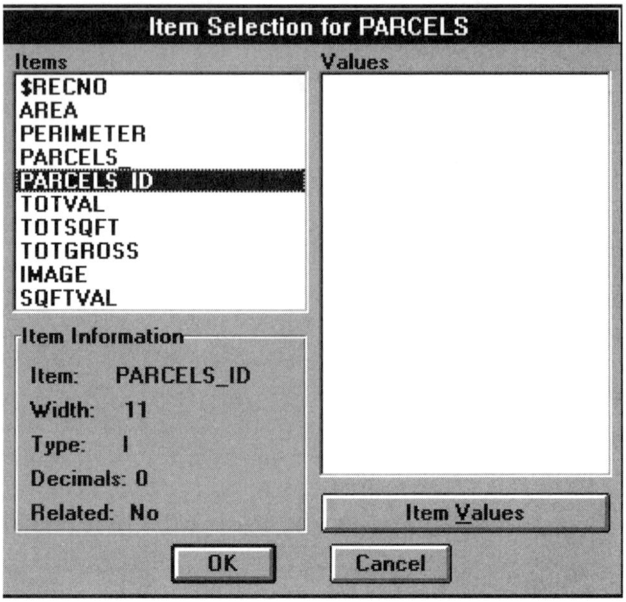

Select PARCELS_ID from the Items list.

`Relate method? (Link/Ordered/Table/<LInear>): ENTER`

ArcCAD now has made a temporary link between the PARCEL's polygon theme and the LOTINFO record theme. To show the relationship between the two tables, execute the Wbrowse command found on the Attributes Datafiles menu.

You should see a series of items preceded by a # symbol. These items are the items from the related LOTINFO theme. The FAT should look like the following figure.

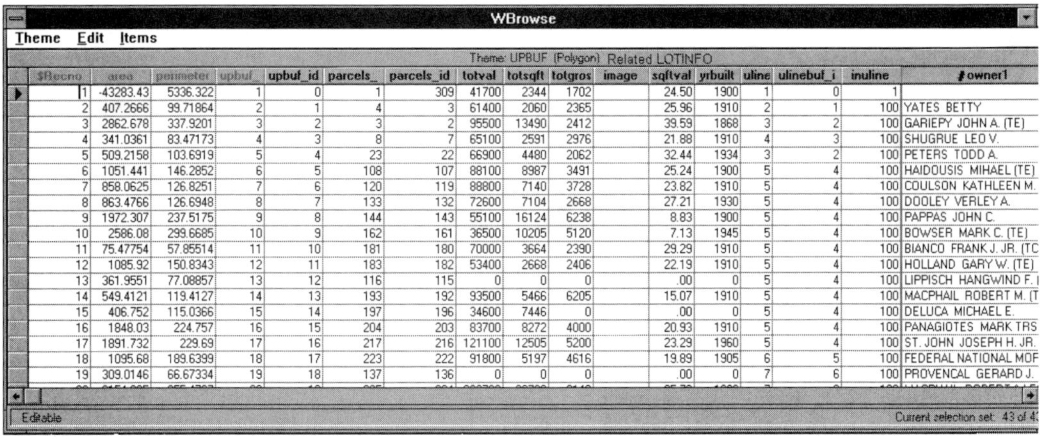
WBrowse

Theme Edit Items

Theme: UPBUF (Polygon) Related LOTINFO

$Recno	area	perimeter	upbuf_	upbuf_id	parcels_	parcels_id	totval	totsqft	totgros	image	sqftval	yrbuilt	uline	ulinebuf_i	inuline	#owner1
1	-43283.43	5336.322	1	0	1	309	41700	2344	1702		24.50	1900	1	0	1	
2	407.2666	99.71864	2	1	4	3	61400	2060	2365		25.96	1910	2	1	100	YATES BETTY
3	2862.678	337.9201	3	2	3	2	95500	13490	2412		39.59	1868	3	2	100	GARIEPY JOHN A. (TE)
4	341.0361	83.47173	4	3	8	7	65100	2591	2976		21.88	1910	4	3	100	SHUGRUE LEO V.
5	509.2158	103.6919	5	4	23	22	66900	4480	2062		32.44	1934	3	2	100	PETERS TODD A.
6	1051.441	146.2852	6	5	108	107	88100	8987	3491		25.24	1900	5	4	100	HAIDOUSIS MIHAEL (TE)
7	858.0625	126.8251	7	6	120	119	88800	7140	3728		23.82	1910	5	4	100	COULSON KATHLEEN M.
8	863.4766	126.6948	8	7	133	132	72600	7104	2668		27.21	1930	5	4	100	DOOLEY VERLEY A.
9	1972.307	237.5175	9	8	144	143	55100	16124	6238		8.83	1900	5	4	100	PAPPAS JOHN C.
10	2586.08	299.6685	10	9	162	161	36500	10205	5120		7.13	1945	5	4	100	BOWSER MARK C. (TE)
11	75.47754	57.85514	11	10	181	180	70000	3664	2390		29.29	1910	5	4	100	BIANCO FRANK J. JR. (TC
12	1085.92	150.8343	12	11	183	182	53400	2668	2406		22.19	1910	5	4	100	HOLLAND GARY W. (TE)
13	361.9551	77.08857	13	12	116	115	0	0	0		.00	0	5	4	100	LIPPISCH HANGWIND F.
14	549.4121	119.4127	14	13	193	192	93500	5466	6205		15.07	1910	5	4	100	MACPHAIL ROBERT M. (T
15	406.752	115.0366	15	14	197	196	34600	7446	0		.00	0	5	4	100	DELUCA MICHAEL E.
16	1848.03	224.757	16	15	204	203	83700	8272	4000		20.93	1910	5	4	100	PANAGIOTES MARK TRS
17	1891.732	229.69	17	16	217	216	121100	12505	5200		23.29	1960	5	4	100	ST. JOHN JOSEPH H. JR.
18	1095.68	189.6399	18	17	223	222	91800	5197	4616		19.89	1905	6	5	100	FEDERAL NATIONAL MOF
19	309.0146	66.67334	19	18	137	136	0	0	0		.00	0	7	6	100	PROVENCAL GERARD J.

Editable

Current selection set: 43 of 4

The PARCELS FAT with related items.

To exit the Wbrowse command, select Exit from the dialog box's Theme menu.

A relate remains in effect until either the drawing is exited or the relate is turned off.

The Ddrelate command found on the Attributes menu may simplify the process of creating relates. For more information on Ddrelate, refer to the *ArcCAD Command Reference.*

Lookup Tables

Another use for record themes are as lookup tables. These database files allow ArcCAD features to be displayed based on symbol numbers, labeled based on a value, or buffered based on a lookup distance. Lookup tables generally contain only two items. The first item defines a link to a FAT, and the second item is either a symbol number, buffer distance, or label string.

The Item names and order of a lookup table are very important. The first item of the lookup table must be the link item—an item with a name and definition that matches that of an item in the corresponding FAT. The second item has one of three names: SYMBOL, LABEL, or DIST. The SYMBOL and DIST items must be numeric, and the LABEL item may be either numeric or alphabetical.

To create a lookup table, you simply define a record theme, then add items and records to the theme.

The following figure shows the lookup table to be created for use with the STREETS theme.

The CLASS lookup table.

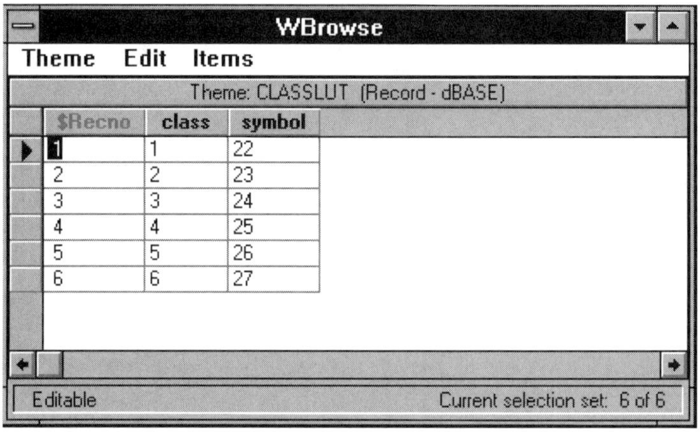

To create the table shown, you must first define a record theme. You will use a new method to create the theme for the STREET lookup table.

Choose Ddtheme from the Themes menu. The Ddtheme dialog box is shown below.

The Ddtheme dialog box for the CLASSLUT theme.

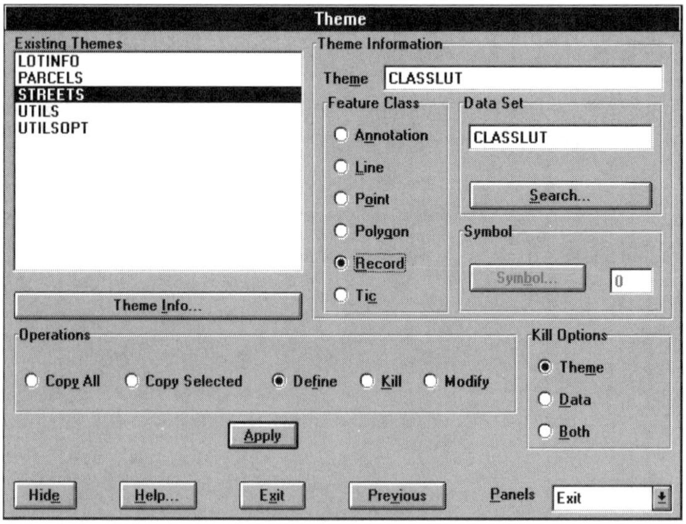

The Ddtheme command allows access to all commands found on the Themes menu through a dialog box.

To create the CLASSLUT record theme, fill out the dialog box as shown and select Apply. Once Apply has been selected the dialog box will return for further operations. You can now select Exit from the dialog box.

With the CLASSLUT theme defined, use the Additem command from the Attributes menu to add two items to the theme. The two items to be added are the link item from the STREETS coverage and the line symbol number for the street centerlines.

The first item to add to the CLASSLUT theme is named CLASS. The second item will be SYMBOL. Follow the input below to add these items.

```
Command: ADDITEM
Theme name (?/<streets>): CLASSLUT
Item name (?): CLASS
Item width: 1
Item type?  (Character/Date/<Numeric>): C
Start item <last>: ENTER

Command: ADDITEM
Theme name (?/<CLASSLUT>): ENTER
Item name (?): SYMBOL
Item width: 3
Item type?  (Character/Date/<Numeric>): ENTER
Item decimal places <0>: ENTER
Start item <last>: ENTER
```

With these items added to the CLASSLUT, records must be added defining the class value and the display symbol. Add records to the CLASSLUT theme, with the Addrec command on the Attributes menu under Records.

```
Command: ADDREC
Theme name (?/<CLASSLUT>): ENTER
```

Continue adding records to the CLASSLUT as shown in the following table. When all records have been added, use the Wbrowse command on the Attributes menu to confirm the data.

To demonstrate the use of a lookup table, use the Arclines command found on the Display pulldown menu under Lines.

```
Command: ARCLINES
Theme name (?/<CLASSLUT>): STREETS
Display using? (Properties/Item/<Symbol>): I
Item (?): CLASS
Optional Lookup table (?/<none>): CLASSLUT
```

The STREET Centerlines displayed using the CLASSLUT lookup table.

Additional examples of lookup tables can be found in the Case Studies at the end of this book. For more information on the use of lookup tables, see the *ArcCAD User's Guide*.

External Database Programs and ArcCAD

ArcCAD uses industry standard dBASE III (.DBF) format files to store its attribute tables and record themes. This allows the data to be edited using any database or spreadsheet package that can read and write .DBF files. Although no other software is necessary, programs such as Microsoft's Excel or Access can simplify the process of data entry and manipulation.

When using another software package to modify ArcCAD tables, it is important to maintain strict dBASE file structure. dBASE requires that item values begin with a letter, contain nothing but letters, numbers, and the underscore character, and be 10 characters or fewer in length.

Non-dBASE Tables and ArcCAD

With version 11.4, ArcCAD has the added the ability to connect to non-dBASE format database tables. To understand how ArcCAD reads these non-dBASE files, you need to know what structured query language (SQL) and the AutoCAD SQL extension are. SQL was developed as a standard method for reading and writing database information. Over the years, SQL has become a standard for many database software manufacturers. Using SQL and a specific driver, the same command syntax may be used to read multiple database formats. In release 12, AutoCAD gained the ability to use these standard commands to read and write to SQL databases and provided several drivers, allowing access to databases created with applications such as Paradox, Oracle, and Informix. Release 11.4 of ArcCAD is taking advantage of these new AutoCAD features to also read SQL databases.

ArcCAD implements SQL functionality by allowing SQL themes to be defined. SQL themes are record themes that use SQL databases as the GIS data set. To define an SQL theme, the Defthm command on the Themes menu is used. The SQL theme is given a theme name, theme type, and GIS data set name. In addition, the SQL theme also specifies a database management system (DBMS), and optionally a login name, password, and SQL expression. The login name and password are only necessary if the source database has them assigned. The SQL expression specified at the bottom of the dialog box allows for an expression to select specific records from the source database. These selected records become the record theme in ArcCAD.

Before defining SQL themes in ArcCAD, the AutoCAD ASI interface must be properly configured. For more information on configuring AutoCAD's ASI interface, refer to Chapter 7 in the *AutoCAD Installation Guide* and Chapter 13 in the *AutoCAD User's Guide.*

Open Database Connectivity

When running in Microsoft Windows, ArcCAD has the ability to work with any Open Database Connectivity (ODBC) compatible database using an SQL theme. ODBC is a set of standards that Microsoft Windows uses to talk to databases.

Accessing Nonstandard Data

So far you have seen how ArcCAD handles accessing both dBASE format files and industry standard SQL databases. In this section you will learn how to access nonstandard database information in ArcCAD.

The LoadDB command found on the Attributes pulldown menu under Datafiles allows a comma- or space-delimited file to be read into a dBASE format database. The database must exist before the command can be used.

The sample project data contains a comma-delimited ASCII file named PARCELDT.TXT. This file could have been created from almost any database product currently on the market. By exporting the data to a comma-delimited format, you can import it into an ArcCAD database.

> Before the file can be imported, the record theme that will hold the data must be created. Use the Defthm command found on the Themes menu to create a record theme named PARDAT. Fill out the dialog box as shown below.

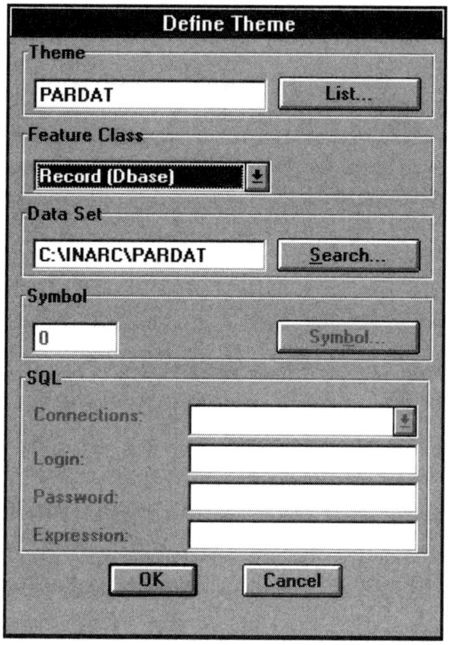

The Defthm dialog box for PARDAT.

With the theme defined, items must be added that match the columns in the comma-delimitted file. If there are four columns, each 20 characters wide in the ASCII file, then four items, each with a width of 20, must be defined on the RECORD theme. To add items to the theme use the Additem command on the Attributes menu under Items.

```
Command: ADDITEM
Theme name (?/<PARDAT>): ENTER
Item name (?): PARCEL_ID
Item width: 11
Item type?  (Character/Date/<Numeric>):
Item decimal places <0>: ENTER
Start item <last>: ENTER

Command: ADDITEM
Theme name (?/<PARDAT>): ENTER
Item name (?): MAPBLOKLOT
Item width: 30
Item type?  (Character/Date/<Numeric>): C
Start item <last>: ENTER
```

```
Command: ADDITEM
Theme name (?/<PARDAT>): ENTER
Item name (?): YRBUILT
Item width: 6
Item type? (Character/Date/<Numeric>):
Item decimal places <0>: ENTER
Start item <last>: ENTER

Command: ADDITEM
Theme name (?/<PARDAT>): ENTER
Item name (?): OCCUP
Item width: 1
Item type? (Character/Date/<Numeric>): C
Start item <last>: ENTER
```

To confirm that all of the items have been added correctly, use the Items command on the Attributes menu. This command will list all of the items and their definitions.

The next step is to import the values from the ASCII file into the database. Use the LoadDB command on the Attributes menu. The syntax for LoadDB is shown below.

```
Command: LOADDB
Theme name (?/<PARDAT>): ENTER
Input file (?): PARCELDT.TXT
Input format (Fixed/Long/<Delimited>): ENTER
747 records loaded.
747 records total.
```

Evaluate the results of the LoadDB command with the Wbrowse command. The database table is shown in the following figure.

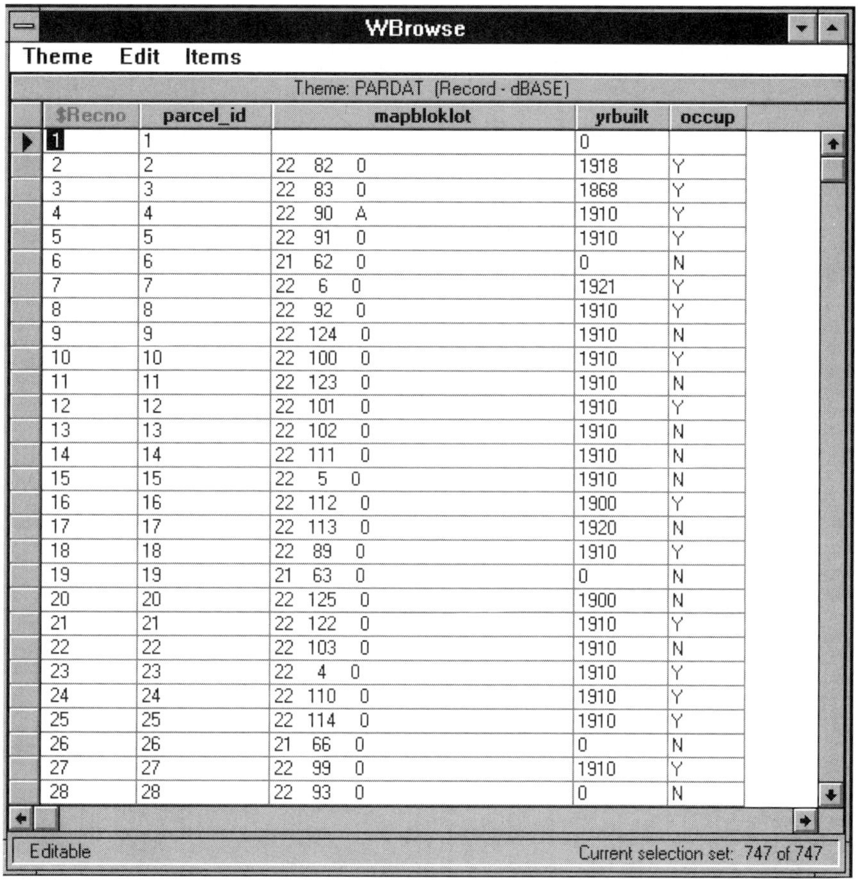

$Recno	parcel_id	mapbloklot	yrbuilt	occup
1	1		0	
2	2	22 82 0	1918	Y
3	3	22 83 0	1868	Y
4	4	22 90 A	1910	Y
5	5	22 91 0	1910	Y
6	6	21 62 0	0	N
7	7	22 6 0	1921	Y
8	8	22 92 0	1910	Y
9	9	22 124 0	1910	N
10	10	22 100 0	1910	Y
11	11	22 123 0	1910	N
12	12	22 101 0	1910	Y
13	13	22 102 0	1910	N
14	14	22 111 0	1910	N
15	15	22 5 0	1910	N
16	16	22 112 0	1900	Y
17	17	22 113 0	1920	N
18	18	22 89 0	1910	Y
19	19	21 63 0	0	N
20	20	22 125 0	1900	N
21	21	22 122 0	1910	Y
22	22	22 103 0	1910	N
23	23	22 4 0	1910	Y
24	24	22 110 0	1910	Y
25	25	22 114 0	1910	Y
26	26	21 66 0	0	N
27	27	22 99 0	1910	Y
28	28	22 93 0	0	N

The PARDAT record theme.

Using what you learned earlier, try relating the PARDAT record theme to the PARCELS coverage. The relate item is the PARCEL_ID item.

Review

Record themes are a powerful feature of ArcCAD. Record themes are used to create optional property tables and lookup tables. They provide a method for maintaining data independent of the spatial data and allow access to most industry standard database formats.

In this chapter you have learned to create record themes and to link them both temporarily and permanently to FATs. You have seen how to access existing dBASE format files and how to create SQL themes to access non-dBASE format files. The LoadDB command was used to take data access one step further and allow access to any data that can be written to a comma- or space-delimited file. In the next several chapters you will continue using record themes to query, display, and analyze the sample GIS data.

Part II.

Analyzing Geographic Data

Chapter

Preparing for Spatial Analysis

Biting Off the Right Amount To Chew

This chapter introduces you to the commands necessary to reduce large data sets into managable study areas. These commands allow you to select specific features to perform analysis on, to select groups of features based on spatial and tabular queries, and to refine data forms for specific analysis.

The commands you will learn about include the following:

- Clip command
- Extract command
- Eliminate command
- Nodepoint command
- Near command
- Xselect command

Setup

To prepare for this chapter,

Start AutoCAD and open the INARC8 drawing file.

Load ArcCAD if it is not set to load automatically. The drawing contains all themes created earlier, and a clip polygon has been drawn.

Managing Large Data Sets

One of ArcCAD's greatest features is that it allows you to work with only the data you need. You are never forced to work with an entire data set if you are only interested in a small portion of the data. This allows ArcCAD to be used with extremely large GIS projects. The commands you will work with in this session will allow you to clip out small portions of larger data sets.

ArcCAD, as an ADS application running inside AutoCAD, requires a considerable amount of a PC's resources to operate efficiently. With very large data sets, even the best equipped PC will slow considerably. This, coupled with the fact that ArcCAD puts certain limits on coverage sizes (see the *ArcCAD User's Guide* for limitations), means you must be able to effectively manage large data sets. This section introduces you to commands that help to accomplish this.

Clipping Themes Down to Size

The first command you will use is Clip. The Clip command takes a point, line, or polygon theme as its input and a polygon theme as the output theme. It then cuts all features in the input theme to the new output theme.

> The BLDGS theme in the project data contains the buildings for a much larger area than you need. To view the buildings in the current theme, execute the Setext command found on the Display menu. At the theme name prompt enter BLDGS and press Enter twice. Next, select the Polys command found on the Display pulldown menu under Polygons and follow the input below.

```
Command: POLYS
Theme name (?/<TEST>): BLDGS
Line symbol number (0-100) (?/): 2
```

The drawing should look like the following figure.

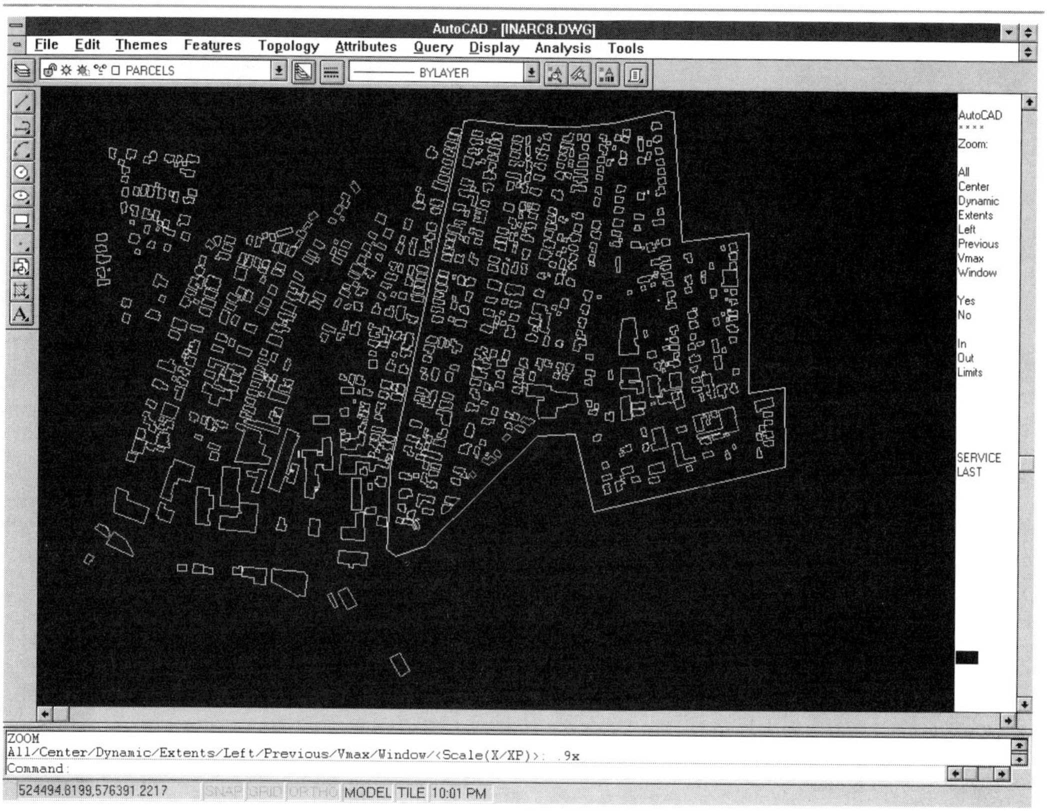

INARC8 drawing with buildings displayed.

In this first exercise you will clip the BLDGS theme to the limits of the other data. To do this, a clip theme must first be created. The BLDGS will then be clipped to this theme and a new theme named BLDGSCLP will be created.

First, define a polygon theme named CLIP using Defthm found on the Themes menu. The dialog box for this theme is shown below.

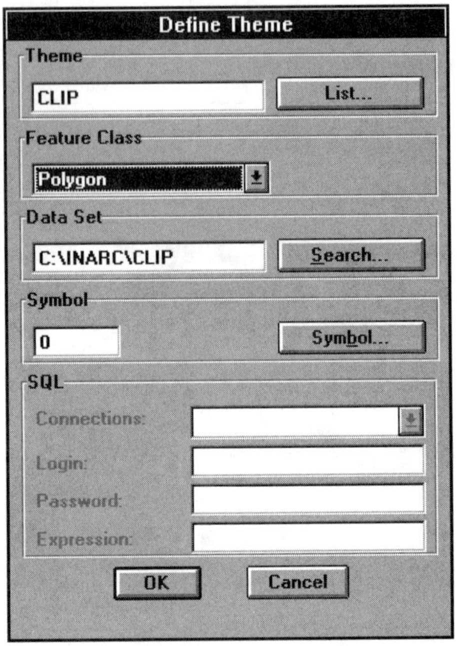

Defthm dialog box for the CLIP theme.

With the theme now defined, use the Addfeat command and select the white polygon. The input for Addfeat is shown below.

```
Command: ADDFEAT
Theme name (?/<CLIP>): ENTER
Select objects: Select the white polyline
Select objects: ENTER
Optional property table (?/<none>): ENTER
1 written to C:\INARC\CLIP, 0 duplicated, 0 ignored.
```

Select Clean Now from the Update Topology dialog box.

```
Output theme (?/<CLIP>): ENTER
Dangle length <default>: ENTER
Fuzzy tolerance <compute>: ENTER
Cleaning CLIP.
Sorting...
Intersecting...
Assembling polygons...
  Sorting input file...
  Sorting label file...
  Processing...
  Assigning final IDs...
```

```
  Writing arc file...
  Generating polygon report...
Creating Polygon attribute table.
Dropping entity-feature links...
```

With the CLIP theme now defined, execute the Clip command found on the Analysis menu and follow the input.

```
Command: CLIP
Input theme (?/<CLIP>): BLDGS
Clip theme (?): CLIP
Clip method? (Net/<Polygon>): ENTER
Output theme (?): BLDGSCLP
Theme BLDGSCLP is not defined.
Defining BLDGSCLP...
Fuzzy tolerance <compute>: ENTER
Clipping BLDGS with CLIP to create BLDGSCLP.
Sorting...
Intersecting...
Assembling polygons...
  Sorting input file...
  Processing...
  Assigning final IDs...
  Writing arc file...
  Generating polygon report...
Labeling...
Creating Polygon attribute table.
```

The clip methods available are Net or Polygon and are defined as follows:

Polygon: For polygon themes only the polygon attribute table will be updated.

Net: For polygon themes, both the polygon attribute table and the arc attribute table will be updated.

With the BLDGS coverage clipped, erase all entities from the screen. Reset the limits of the screen by executing the Setext command from the Display menu. Display the new BLDGSCLP theme into the drawing by choosing Polys from the Display menu. The input is shown below.

```
Command: POLYS
Theme name (?/<BLDGSCLP>): ENTER
Line symbol number (0-100) (?/<0>): 2
```

The drawing should look like the following figure.

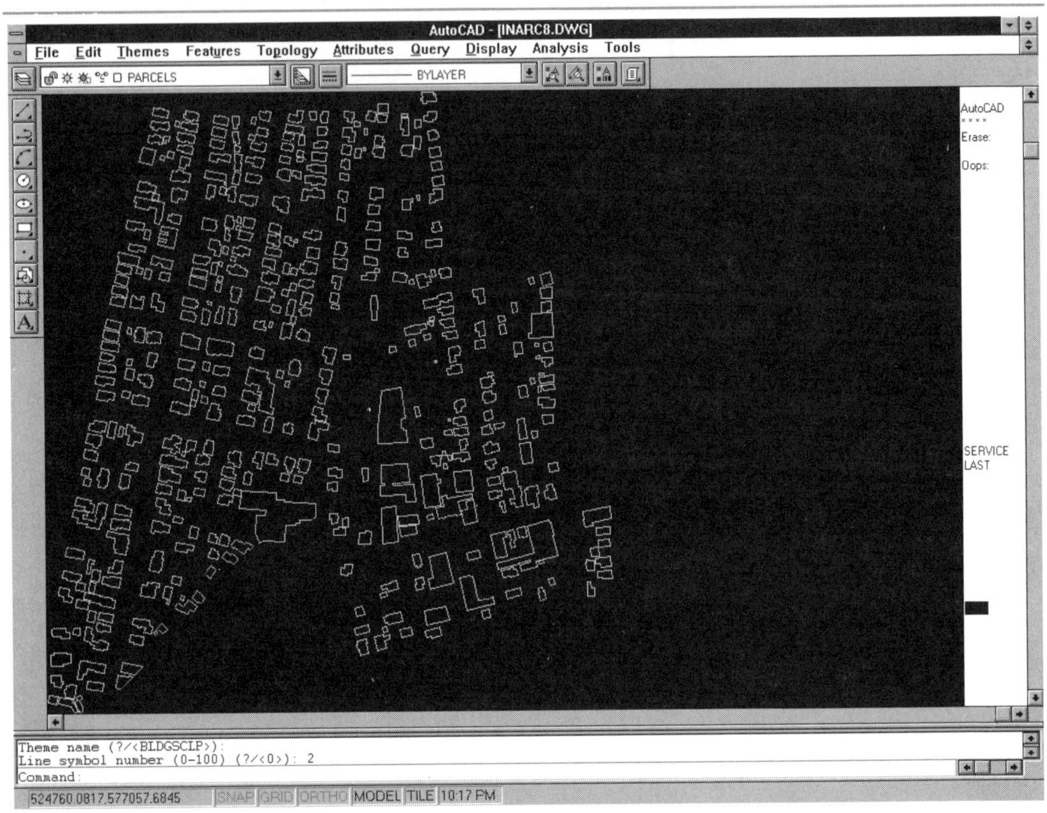

INARC8 drawing with clipped buildings.

You have now completed the process of clipping a subset of the buildings out of the original coverage. The Update command found on the Analysis menu can be used with the Clip command to allow any changes in the BLDGSCLP theme to be written back to the original BLDGS theme. For more information on the Update command, refer to the *ArcCAD User's Guide* or *Command Reference.*

Extracting Selected Features

You have used the Clip command to pull a set of features from one theme and place them into a new theme. In this section you will use a similar command that will extract the features based on a database query as opposed to a spatial selection.

The Extract command, found on the Analysis menu, copies all features in the current selection set and places them in a new coverage. A selection set is a set of features that meets a specific set of spatial or tabular criteria. In this exercise you will extract one of the street centerlines. Later this centerline will be used as the input for a buffer. Since the Buffer command works only on entire themes, it is necessary to first extract the required centerline.

The street of interest is Cedar Street. To extract this street from the STREETS theme, you must first select the feature.

> Execute the Reselect command found on the Query menu and follow the imput.

```
Command: RESELECT
Theme name (?/<BLDGSCLP>): STREETS
Reselect by?
  (Circle/ENtity/Expression/Polygon/Window): E
```

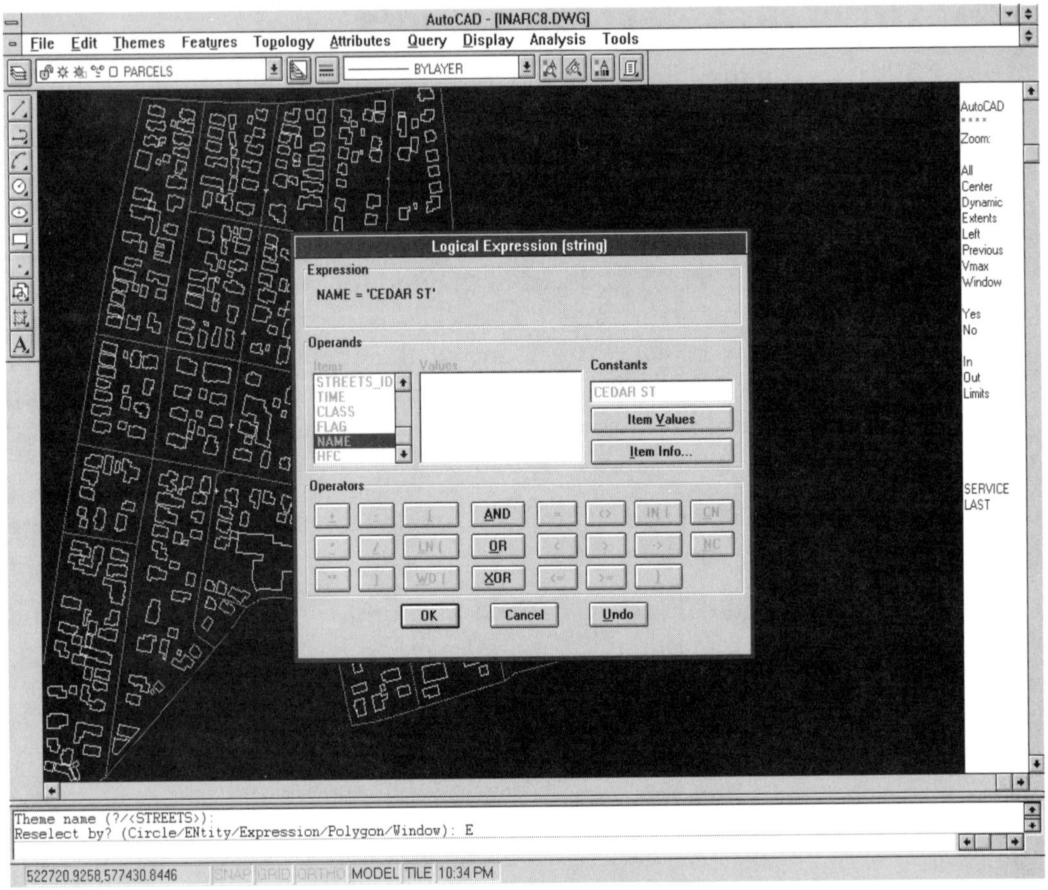

The Logical Expression dialog box for STREETS.

In the Logical Expression dialog box, select NAME from the Item list, choose = from the operators, and type `CEDAR ST` into the value box. Be sure to press ENTER after typing the value. Select OK from the dialog box. ArcCAD should return the following response.

```
" NAME = 'CEDAR ST'" C:\INARC\STREETS arcs : 3 of 70
  selected.
```

To display this selection set, choose Arclines from the Display pulldown menu under Lines. The input is shown below.

```
Command: ARCLINES
Theme name (?/<STREETS>): ENTER
Display using? (Properties/Item/<Symbol>): ENTER
Line symbol number (0-100) (?/<0>): 1
```

NOTE: *Display commands only work on the selected set.*

The drawing should now look like the following figure.

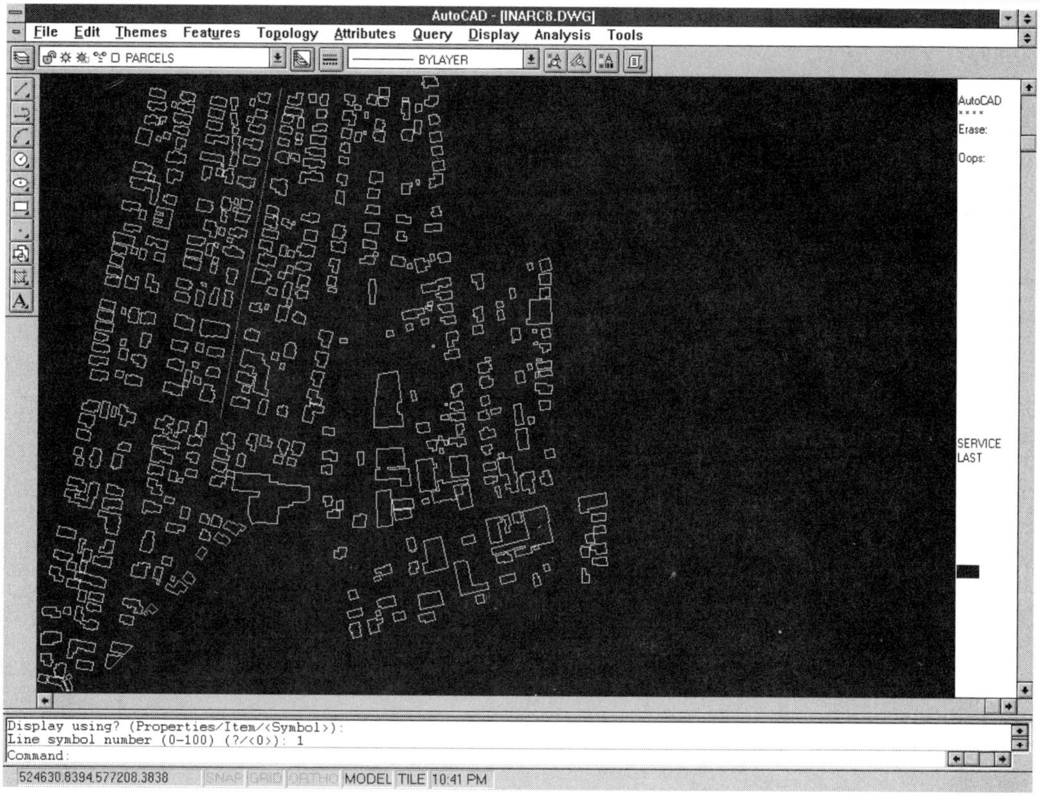

The INARC8 drawing with Cedar Street displayed.

The next step is to extract Cedar Street to a new coverage. Choose Extract from the Analysis menu. The input for the Extract command is shown below.

```
Command: EXTRACT
Input theme (?/<STREETS>): ENTER
Output theme (?): CEDAR
```

```
Theme CEDAR is not defined.
```

Select OK from the Define Theme dialog box.

```
Defining CEDAR...
Extracting features from STREETS to create CEDAR.
3 of 70 records selected.
Extracting lines...
Number of Arcs in Coverage...      70
Number of Arcs Reselected....       3
Creating Line attribute table.
```

Cedar Street has now been copied from the STREETS theme to a new theme named CEDAR. The original features in the STREETS theme have not been altered.

To view the new theme, select Wbrowse from the Attributes menu. At the theme name prompt enter CEDAR and press return. The new theme should look like the following figure.

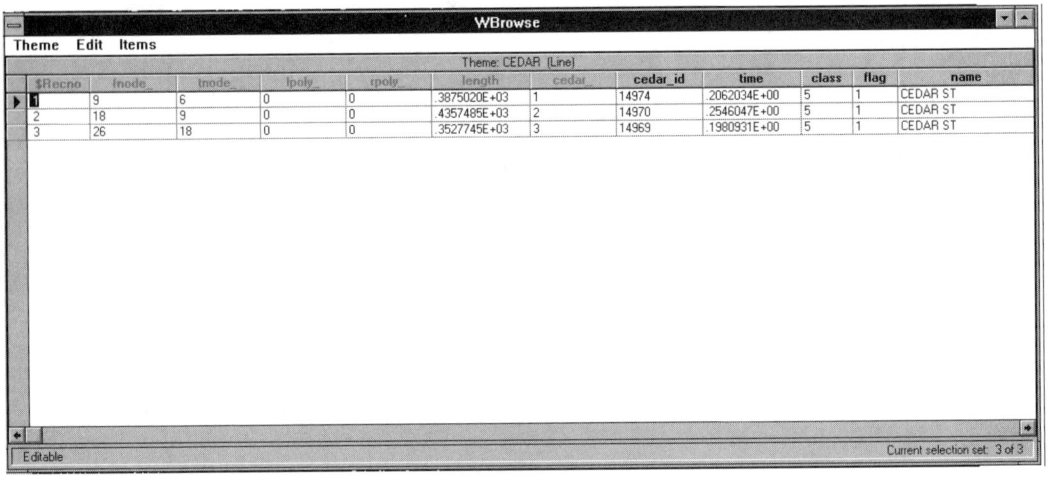
WBrowse

Theme Edit Items

Theme: CEDAR (Line)

$Recno	fnode_	tnode_	lpoly_	rpoly_	length	cedar_	cedar_id	time	class	flag	name
1	9	6	0	0	.3875020E+03	1	14974	.2062034E+00	5	1	CEDAR ST
2	18	9	0	0	.4357485E+03	2	14970	.2546047E+00	5	1	CEDAR ST
3	26	18	0	0	.3527745E+03	3	14969	.1980931E+00	5	1	CEDAR ST

Editable

Current selection set: 3 of 3

The CEDAR FAT.

The Extract command has copied the selected features from the STREETS theme to a new theme named CEDAR. This command has the same outcome as the Clip command, with one exception. The Extract command takes whole features, whereas the Clip command clips a feature at the clip boundary, leaving partial polygons and line segments in the output theme.

Removing Unneeded Data

You have seen several commands for reducing data sets by copying only the needed features to new themes. In this example you will use a command that will reduce a data set by eliminating unneeded information. The example will evaluate the LANDUSE polygons based on size. The polygons with less than a user-specified area will be deleted from the theme. This command is excellent for removing the long, thin sliver polygons that can be created when overlaying polygon themes or using the Mapjoin or Append command.

Before eliminating the sliver polygons in the LANDUSE theme, use the AutoCAD Erase command to clear the screen.

To eliminate features from a data set you must first select the features as you did with the Extract command.

To select all polygons with an area less than 2500 square feet in the LANDUSE theme, execute the Reselect command from the Query menu and follow the input below.

```
Command: RESELECT
Theme name (?/<LANDUSE>): ENTER
Reselect by?
  (Circle/ENtity/Expression/Polygon/Window): E
```

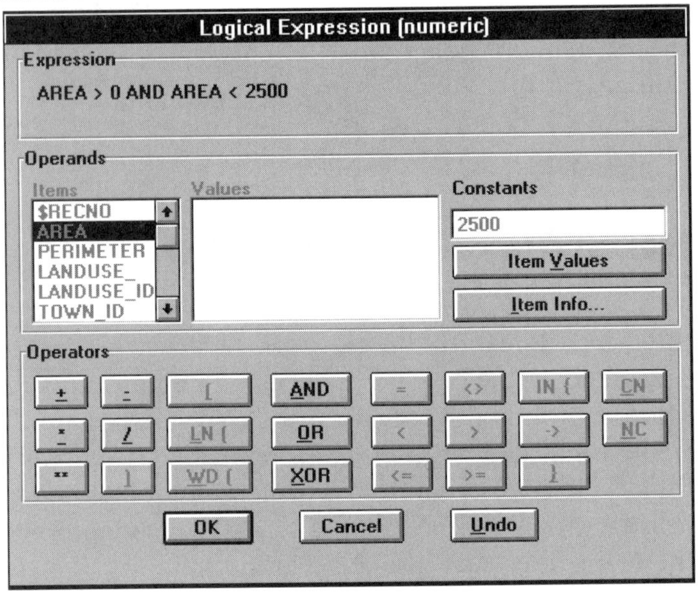

The Logical Expression dialog box for LANDUSE.

```
" AREA > 0 AND AREA < 2500" C:\INARC\LANDUSE polys :
  2 of 10 selected.
```

ArcCAD selects two polygons. These two polygons can now be eliminated from the theme using the Eliminate command.

From the Analysis menu, select Eliminate and follow the input below.

```
Command: ELIMINATE
Input theme (?/<LANDUSE>): ENTER
Output theme (?): LANDUSE2
Theme LANDUSE2 is not defined.
```

Select OK from the Define Theme dialog box.

```
Defining LANDUSE2...
Keep edge polygons? (Yes/<No>): ENTER
Eliminating polygons in LANDUSE to create LANDUSE2.
2 of 10 records selected.
Number of Polygons (Input,Output) =       10         8
Number of Arcs     (Input,Output) =       30        28
Creating Polygon attribute table.
```

"Keep edge polygons? (Yes/<No>)" determines whether boundary polygons may be eliminated or not.

With these two polygons eliminated from the LANDUSE theme, set the LANDUSE layer current and display the polygons in the drawing.

```
Command: LA
LAYER ?/Make/Set/New/ON/OFF/Color/Ltype/Freeze/Thaw/
  LOck/Unlock: S
New current layer <STREETS>: LANDUSE
?/Make/Set/New/ON/OFF/Color/Ltype/Freeze/Thaw/LOck/
  Unlock: ENTER
Command: POLYS
Theme name (?/<LANDUSE2>): ENTER
Line symbol number (0-100) (?/<0>): 4
```

The new LANDUSE polygons should look like those in the following figure.

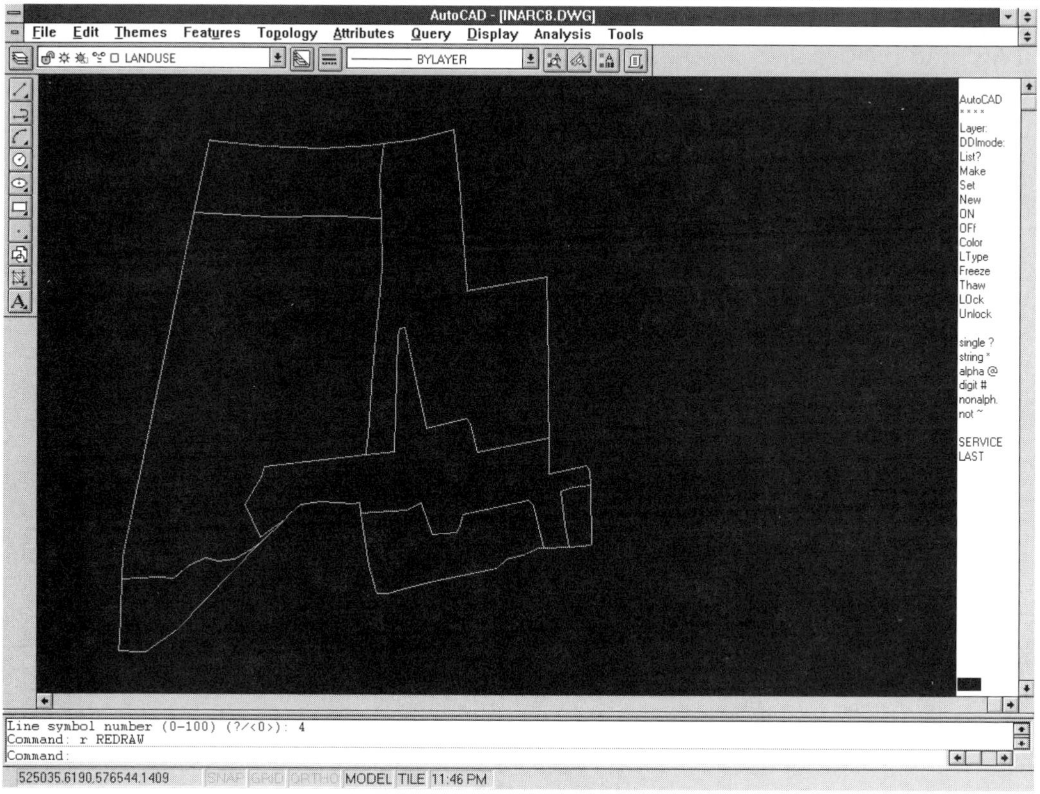

The INARC8 drawing with the new LANDUSE polygons.

Calculating New Data from Existing Themes

In the last section of preparing data you will work with the STREETS theme. The task is to prepare the data so that the analysis can determine how many street lights are within 50 feet of a road intersection. In order to make this analysis, you must have an ArcCAD feature representing each road intersection. This could be done using AutoCAD to plot a point at each intersection. A theme could be defined, and the new points could be written to it. However, this method not only does not use the spatial abilities of ArcCAD, but may also require a lot of work, considering the project area could be an entire city.

Using ArcCAD, you can create a point theme from the *from* and *to* nodes in the STREETS theme. This will provide you with a point feature at each of the street intersections. The next step will be to extract the internal ID of the line feature closest to each of these new points. This allows you to relate the STREETS FAT to the INTERSECTION POINTS FAT. The related data will then identify the street name of each intersection point.

The concepts discussed in this section are not simple. Working through the sample exercises should help to clarify this information. The two commands used in this section are the Nodepoint and Near commands from the Analysis menu. Although these commands are used on street centerlines in this example, they are powerful commands and have many uses. Please keep an open mind while working through this exercise and think of ways to apply them within your environment.

The Nodepoint command will create a point theme based on all of the *from* and *to* nodes of the selected line or polygon theme.

First, use the AutoCAD Erase command to clear the screen.

Set the STREETS layer current, clear the current selection set, and use the Arclines command to display the street centerlines.

```
Command: LA
LAYER ?/Make/Set/New/ON/OFF/Color/Ltype/Freeze/Thaw/
  LOck/Unlock: S
New current layer <LANDUSE>: STREETS
?/Make/Set/New/ON/OFF/Color/Ltype/Freeze/Thaw/LOck/
  Unlock: ENTER
```

```
Command: CLEARSEL

Command: ARCLINES
Theme name (?/<LANDUSE2>): STREETS
Display using? (Properties/Item/<Symbol>): ENTER
Line symbol number (0-100) (?/<0>): 24
```

The drawing should look as shown in the following figure.

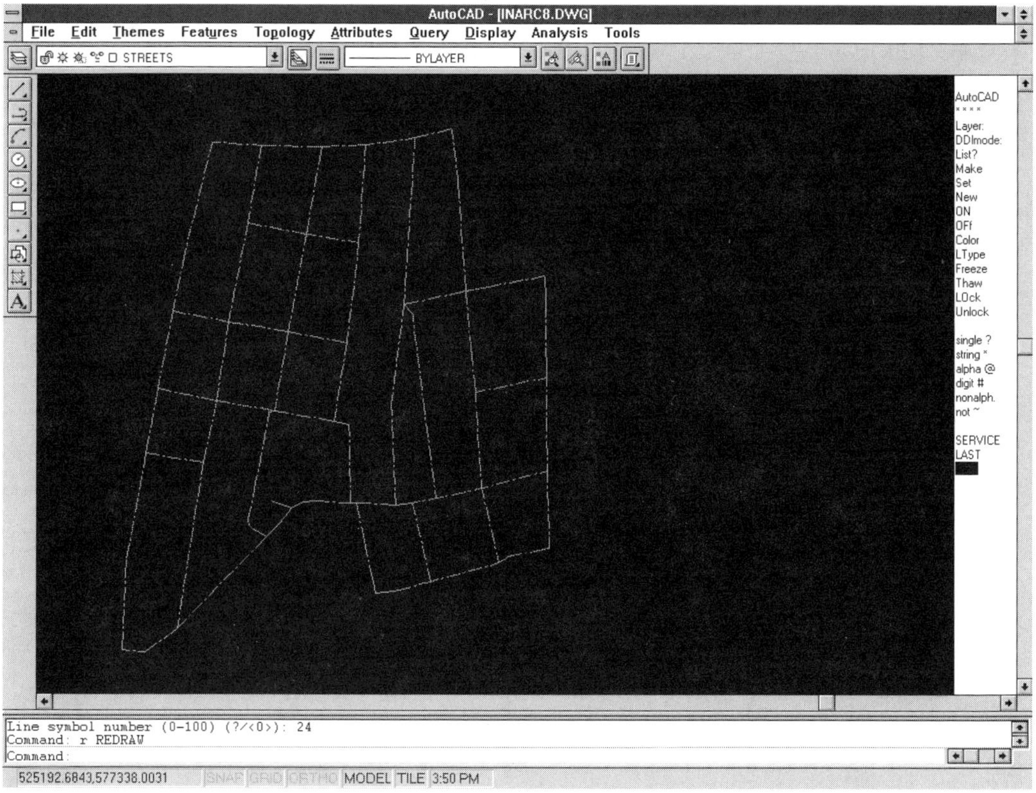

The INARC8 drawing with the STREETS displayed.

With the street centerlines displayed in the drawing, execute the Nodepoint command on the Analysis menu. The input for the Nodepoint command is shown below.

```
Command: NODEPOINT
Input theme (?/<STREETS>): ENTER
Output theme (?): INTER
```

```
Theme INTER is not defined.
```

Select OK from the Define Theme dialog box.

```
Defining INTER...
    50 point(s) generated
Building point attribute table...
Copying INTER to __TEMP_ESRI_THEME...
Building points...
Creating Point attribute table.
```

ArcCAD has now created a new point theme called INTER. The theme contains one point feature for every *from* or *to* node in the STREETS theme.

To view the new point features, make a new layer named INTER and execute the Pointmrk command found on the Display pull-down menu under Points.

```
Command: LA
LAYER ?/Make/Set/New/ON/OFF/Color/Ltype/Freeze/Thaw/
  LOck/Unlock: M
New current layer <STREETS>: INTER
?/Make/Set/New/ON/OFF/Color/Ltype/Freeze/Thaw/LOck/
  Unlock: ENTER

Command: POINTMRK
Theme name (?/<INTER>): ENTER
Display using? (Properties/Item/<Symbol>): ENTER
Marker symbol number (0-100) (?/<0>): 22
```

Quick inspection of the drawing will show that every node of the STREETS theme now has a red square plotted on it from the INTER theme. If there were pseudo nodes in the STREETS theme, those two have received a point. This will cause an error in the analysis to be performed in the next chapter.

The INARC8 drawing with the STREETS and INTER themes displayed.

If you remember, you are identifying every intersection with a street light within 50 feet. With the points plotted on the pseudo nodes, any light within 50 feet of these will be returned as well. To fix this, use the Delfeat command found on the Features menu to remove any point that is not on an intersection. The points to be removed and the input for the Delfeat command are shown in the following figure.

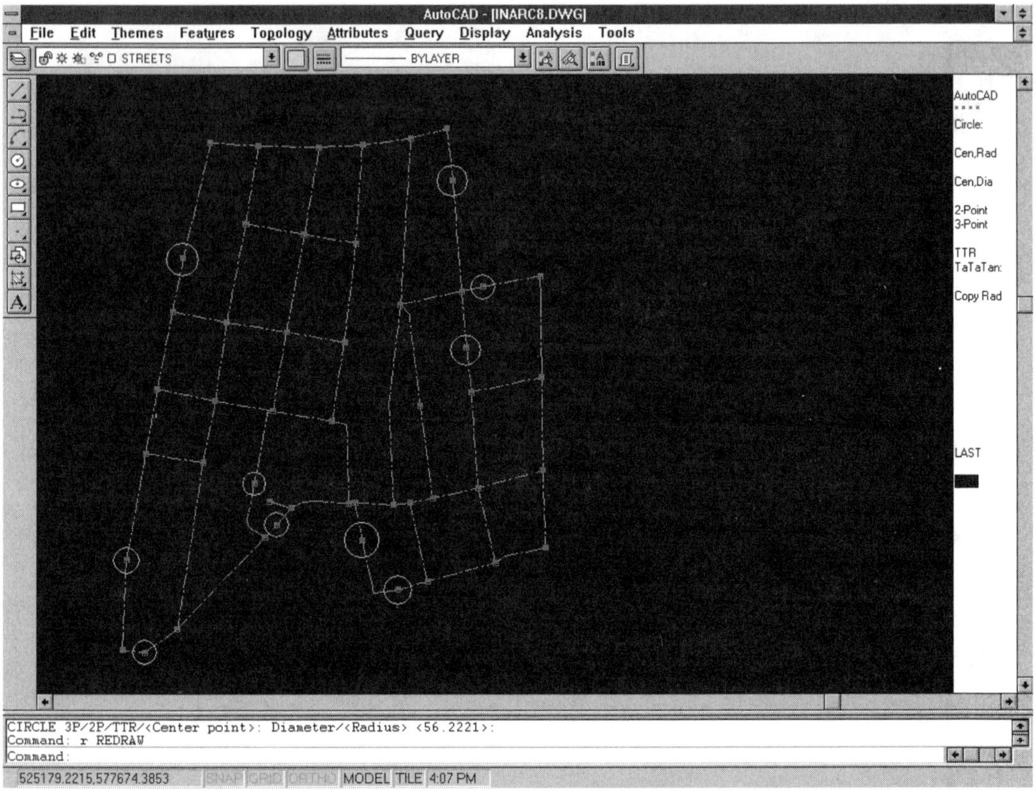

INTER points to be deleted.

```
Command: DELFEAT
Theme name (?/<INTER>): ENTER
Select objects: 1 found
Select objects: ...
10 deleted from C:\INARC\INTER, 0 ignored.
```

Select Build Now from the Update Topology dialog box.

```
Output theme (?/<INTER>): ENTER
Building points...
Creating Point attribute table.
Dropping entity-feature links...
Unlinking...
```

Use the Xselect command found on the Query menu to remove the old points from the drawing. The Xselect command uses property filters to select entities based on a specific property such as color or layer.

```
Command: XSELECT
Property? (Theme or Color/ELevation/Entity/Layer/
  LInetype/THickness): C
Color value: 1
Property?
  (Color/ELevation/Entity/Layer/LInetype/THickness):
41 selected.

Command: ERASE
Select objects: P
41 found
Select objects: ENTER
```

To confirm the deletion of all unwanted points in the INTER theme, use the Pointmrk command found on the Display pulldown menu under POINTS to redisplay the features into the drawing.

```
Command: POINTMRK
Theme name (?/<INTER>): ENTER
Display using? (Properties/Item/<Symbol>): ENTER
Marker symbol number (0-100) (?/<0>): 22
```

If you missed any of the point features, reexecute the Delfeat command and select the points. At the Update Topology dialog box select Build Now and repeat the Erase and Display commands.

Review

The ability to break up large data sets into working pieces is integral to ArcCAD's ability to serve complex GIS needs. You have seen how to break up a data set based on tabular and spatial queries. You have used the Extract and Clip commands and created new data based on existing coverages. The commands used in this chapter are only some of the data manipulation commands provided by ArcCAD. Some of the commands not covered here include Split, Mapjoin, and Append. For more information on how to break coverages up into manageable sizes or rebuild clipped or split coverages, refer to the *ArcCAD User's Guide*.

Chapter 9

Performing Spatial Analysis

Query and Analysis

Spatial analysis is the process of identifying relationships between different sets of geographical data. These relationships are identified by merging geographic and tabular data and then asking questions (querying) of the combined data using a series of tabular and spatial tools.

The concepts you will learn about in this chapter include:

- ❒ Buffering
- ❒ Topological Overlays

Setup

To prepare for this chapter,

Start AutoCAD and open the ROADS drawing file. If you have not already successfully completed the previous chapters, copy the coverages from the C:\INARC\ROADS directory to the C:\INARC directory before opening the drawing.

Load ArcCAD if it is not set to load automatically.

The Analysis

Three separate sets of analysis will be performed on the project data you have compiled and prepared in previous chapters. The first set of analysis will be to determine the approximate cost to widen Cedar Street. The second analysis will be to determine how many street intersections have a light pole within 50 feet, and the last set of analysis will be to compare the number of fire hydrants located in each of the five different zones.

To perform this analysis, you will use all of the concepts you have learned to this point and be introduced to ArcCAD's set of spatial analysis tools.

Buffering: Selecting Features Based on Distance

The Buffer command provides the ability to select all features that are a specific distance from other features. A buffer can be generated for points, lines, and polygon themes. It can be either a constant width or a varying width based on an item in the input theme's feature attribute table (FAT) or lookup table.

The Buffer command always creates a new polygon theme with five items in its FAT. The first four items are the standard items of a polygon theme and the fifth is an item named INSIDE. The INSIDE item identifies the polygon as being inside the buffer (a value of 100) or outside the buffer (a value of 1).

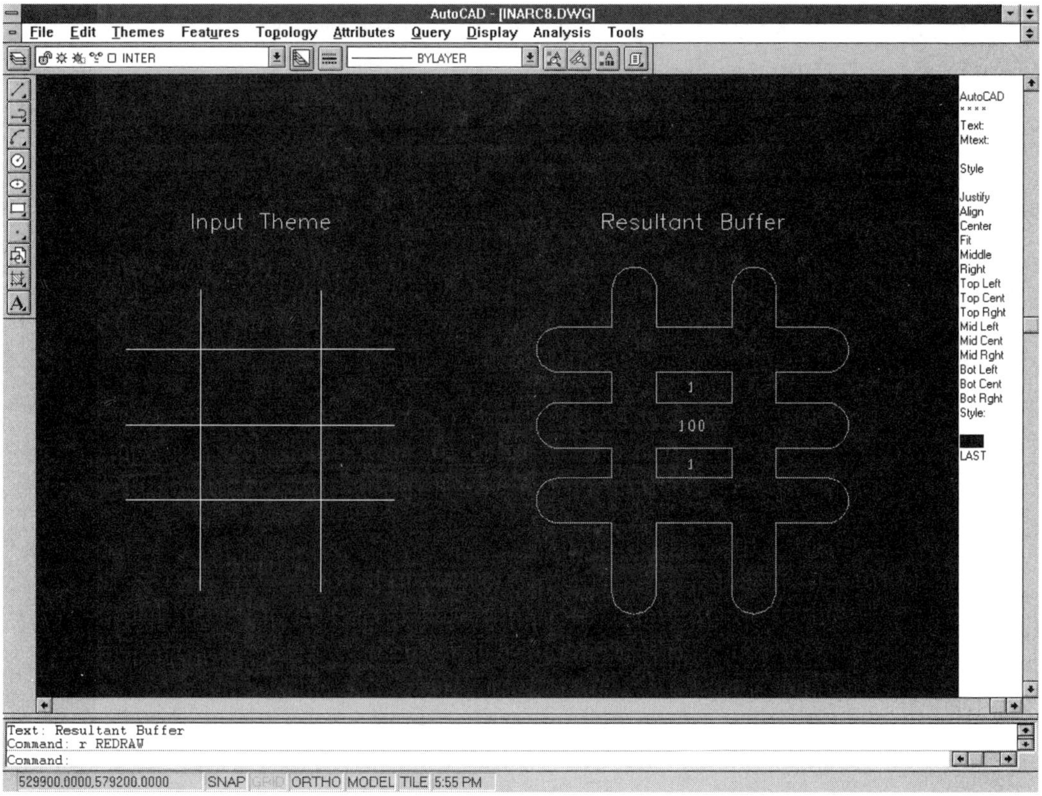

A typical Buffer with INSIDE labeled.

Once the buffer has been created, it can be overlaid with other themes to determine their location relative to the buffer input theme. For example, to find all of the lights within 50 feet of the inter points created earlier, you will create a 50-foot buffer around the inter points. Overlaying the buffer with the lights point theme will identify all lights within 50 feet of an intersection. Later in this chapter you will use the Buffer command to select features spatially.

Topological Overlays: Selecting Features Based on Other Features

The ability to create topological overlays is unique to GIS. The concept of topological overlays is a simple one. It is the ability to combine spatial and tabular information to create new sets of information. For example, in the point in polygon overlay discussed in the previous section, the INTER points overlaid on the buffer theme will return a new set of points. The new set will consist of only the points that fall within a buffer. These new points will contain the attributes of both the input theme, INTER, and the overlay buffer theme.

Overlays may be calculated for point, line, and polygon themes. The possible combinations of overlays are presented below.

Point in polygon: A points theme will be returned containing the attributes of both the input theme and the overlay theme.

Line in polygon: A line theme will be created containing all attributes of both the input theme and the overlay theme.

Polygon in polygon: A polygon theme will be returned containing all attributes of the input theme and the overlay theme.

The three commands used for creating topological overlayers are IDENTITY, ESRI_INTERSECT, and ESRI_UNION.

IDENTITY combines the two themes attribute tables and keeps all features of the input theme and only the features of the overlay theme that intersect with the input theme. An identity overlay can use points, lines, and polygons as its input but only polygons as the overlay theme.

ESRI_INTERSECT combines the two theme's attribute tables and keeps only the features that are common to both. An intersect overlay can use points, lines, and polygons as its input but only polygons as the overlay theme.

ESRI_UNION combines the two themes' attribute tables and keeps all features of both coverages.

For more information on the specifics of topological overlays, refer to the *ArcCAD User's Guide.*

Analyzing a Road Widening Project

The first project you will work through is the analysis of a road widening project. The objectives of this project are to identify the area of land to be taken for the widening and estimate the projected cost of the land. We also want to generate a list of owners' names and addresses of land being considered.

The existing road is approximately 40 feet wide. After the widening the road will be 60 feet wide. Based on this, you need to select all parcels that are within 60 feet of the Cedar Street centerline. To do this you will create a buffer around Cedar Street with a constant width of 60 feet.

All of the themes necessary for the project have already been defined in the ROADS drawing file.

To begin the project, display the polygons for the PARCELS theme using the Polys command found on the Display pulldown menu under Polygons.

```
Command: LA
LAYER ?/Make/Set/New/ON/OFF/Color/Ltype/Freeze/Thaw/
  LOck/Unlock: S
New current layer <0>: PARCELS
?/Make/Set/New/ON/OFF/Color/Ltype/Freeze/Thaw/LOck/
  Unlock: ENTER

Command: POLYS
Theme name (?/<THEME>): PARCELS
Line symbol number (0-100) (?/<0>): 4
```

Next, display Cedar Street from the CEDAR theme using the Arclines command found on the Display pulldown menu under Lines.

```
Command: LA
LAYER
  ?/Make/Set/New/ON/OFF/Color/Ltype/Freeze/Thaw/LOck/
  Unlock: S
New current layer <PARCELS>: STREETS
?/Make/Set/New/ON/OFF/Color/Ltype/Freeze/Thaw/LOck/
  Unlock: ENTER
```

```
Command: ARCLINES
Theme name (?/<PARCELS>): CEDAR
Display using? (Properties/Item/<Symbol>): ENTER
Line symbol number (0-100) (?/<0>): 24
```

The drawing file should look like the following figure.

The ROADS drawing file.

To create the 60-foot buffer use the Buffer command found on the Analysis menu. Notice that in order to create a 60-foot buffer around Cedar Street, 30 feet is entered as the buffer distance. The 30 feet will be applied to both sides of the road, giving a total width of 60 feet.

```
Command: BUFFER
Input theme (?/<CEDAR>): ENTER
```

```
Output theme (?): CEDARBUF
Theme CEDARBUF is not defined.
```

Select OK from the Define Theme dialog box.

```
Defining CEDARBUF...
Buffer using? (Item/<Constant>): ENTER
Buffer distance: 30
Fuzzy tolerance <compute>: ENTER
Buffering...
Sorting...
Intersecting...
Assembling polygons...
  Sorting input file...
  Processing...
  Assigning final IDs...
  Writing arc file...
  Generating polygon report...
Labeling...
Finding inside polygons...
Dissolving...
Number of Polygons (Input,Output) =        6         2
Number of Arcs     (Input,Output) =       12         1
Creating polygon attribute table...
```

Now that the buffer has been created, create a new layer named BUFFER and display it using the Polys command found on the Display pulldown menu under Polygons.

```
Command: LA
LAYER ?/Make/Set/New/ON/OFF/Color/Ltype/Freeze/Thaw/
  LOck/Unlock: M
New current layer <STREETS>: BUFFER
?/Make/Set/New/ON/OFF/Color/Ltype/Freeze/Thaw/LOck/
  Unlock: ENTER

Command: POLYS
Theme name (?/<CEDARBUF>): ENTER
Line symbol number (0-100) (?/<0>): 2
```

The ROADS drawing file with buffer.

To view the FAT for the CEDARBUF theme, execute the Wbrowse command and enter CEDARBUF at the theme name prompt.

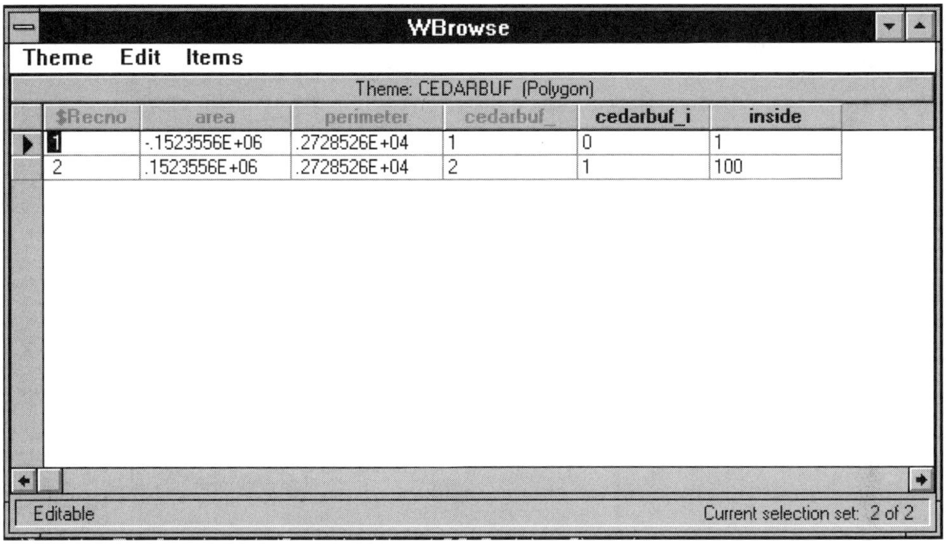

The CEDARBUF FAT.

✔ ***TIP:*** *When working with buffers it is a good habit to change the INSIDE item name as soon as the buffer is created. This is to avoid overwriting the item when overlaying multiple buffer themes.*

To change the INSIDE item name in CEDARBUF, execute the Moditem command found on the Attributes pulldown menu under Items.

```
Command: MODITEM
Theme name (?/<CEDARBUF>): ENTER
Item (?): INSIDE
New item (?/<INSIDE>): INCEDAR
Item width <old>: ENTER
Item type (Character/Date/Numeric) <old>: ENTER
Item decimal places <old>: ENTER
```

Wbrowse will display the results of the Moditem command.

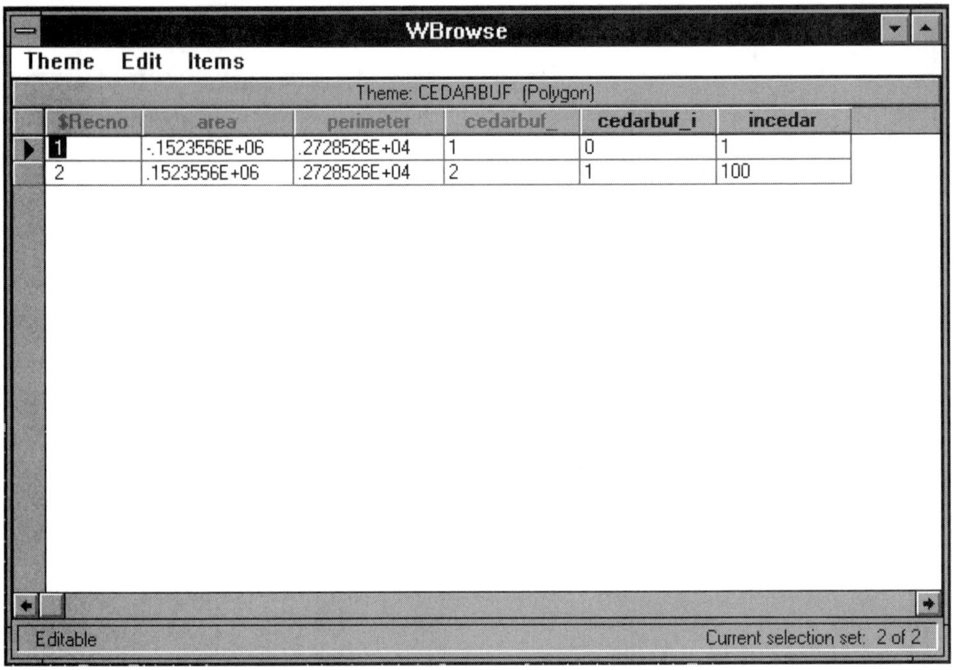

The CEDARBUF FAT.

For more information on modifying items, refer to Chapter 7, "Working with Tables."

As you can see from looking at the drawing, it is obvious which parcels will be affected by the road widening project. The ability to display multiple sets of data simultaneously on the screen is known as *graphic overplot.* Graphic overplot differs from topological overlay in that the overplot does not physically merge the different sets of data or the attribute tables, it simply displays the data. Topological overlay merges the two sets of data and combines the attribute tables.

Now that you have successfully created a buffer theme, this theme can be overlaid with the PARCELS theme to identify the amount of land needed to increase the width of the road.

When creating an overlay, it is important to consider the desired output of the overlay. In many cases more than one command will return the desired information, but typically one is more suitable than the others. For example, to determine the parcel area to take for the road widening project, any of the three commands could be used. However, two of the three will return considerably more data than necessary.

If the ESRI_UNION command is used to calculate the overlay, all polygons of the buffer coverage and all polygons of the parcels coverage will be returned. Since the overlay would also combine the attribute tables, a query would answer the question at hand, but return a considerable amount of unnecessary data.

The Identity command could also be used, but with much the same results as the ESRI_UNION command. By using the ESRI_INTERSECT command to overlay the two themes, you will return the least amount of data necessary to answer the question.

To overlay the two themes, execute the ESRI_INTERSECT command found on the Analysis menu.

```
Command: ESRI_INTERSECT
Input theme (?/<CEDARBUF>): PARCELS
Intersect theme (?): CEDARBUF
Output theme (?): PARCELTK
Theme PARCELTK is not defined.
```

Select OK from the Define Theme dialog box.

```
Defining PARCELTK...
Fuzzy tolerance <compute>: ENTER
Intersecting PARCELS with CEDARBUF to create PARCELTK.
Sorting...
Intersecting...
Assembling polygons...
  Sorting input file...
  Processing...
  Assigning final IDs...
  Writing arc file...
  Generating polygon report...
Labeling...
Creating Polygon attribute table.
Item "AREA" duplicated, Join File version dropped
Item "PERIMETER" duplicated, Join File version dropped
Item "PARCELS_" duplicated, Join File version dropped
Item "AREA" duplicated, Join File version dropped
Item "PERIMETER" duplicated, Join File version dropped
```

```
Item "CEDARBUF_" duplicated, Join File version dropped
```

The ESRI_INTERSECT command has created a new theme whose contents are the polygons that are both part of the CEDARBUF theme and the PARCELS theme.

To view the result of the overlay, execute the Polyshd command found on the Display pulldown menu under Polygons.

```
Command: POLYSHD
Theme name (?/<PARCELTK>): ENTER
Display using? (Properties/Item/<Symbol>): ENTER
Shade symbol number (0-100) (?/<0>): 5
```

The drawing should look like the following figure.

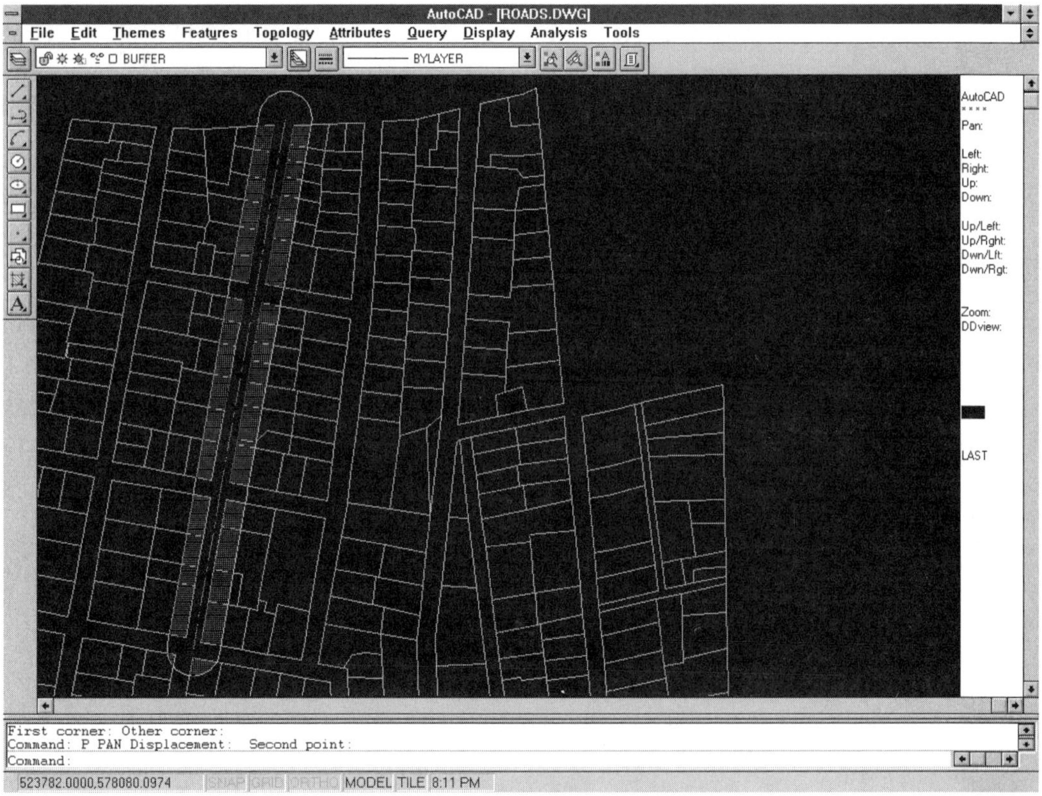

The PARCELTK polygons.

Use the Wbrowse command to view the PARCELTK FAT.

WBrowse

Theme Edit Items

Theme: PARCELTK (Polygon)

$Recno	parceltk_i	parcels_	parcels_id	totval	totsqft	totgross	image	sqftval	cedarbuf_	cedarbuf_i	incedar
1	0	1	309	41700	2344	1702		24.50	1	0	1
2	1	11	10	76100	4151	3219		23.64	2	1	100
3	2	12	11	93200	4134	4560		20.44	2	1	100
4	3	18	17	95500	2741	4904		19.47	2	1	100
5	4	19	18	71900	4640	3706		19.40	2	1	100
6	5	26	25	54300	3057	1944		27.93	2	1	100
7	6	31	30	75500	5399	2752		27.43	2	1	100
8	7	33	32	65700	3594	2120		30.99	2	1	100
9	8	42	41	65100	4513	2064		31.54	2	1	100
10	9	41	40	84500	7124	3296		25.64	2	1	100
11	10	49	48	93200	4503	4691		19.87	2	1	100
12	11	51	50	77800	6663	2875		27.06	2	1	100
13	12	58	57	114400	10061	6135		18.65	2	1	100
14	13	64	63	71200	4848	3128		22.76	2	1	100
15	14	74	73	66700	3560	2444		27.29	2	1	100
16	15	76	75	11500	4391	2602		4.42	2	1	100
17	16	81	80	73700	5617	2430		30.33	2	1	100
18	17	84	83	79600	4540	3516		22.64	2	1	100
19	18	91	90	91400	4483	4372		20.91	2	1	100
20	19	90	89	76600	5637	3118		24.57	2	1	100
21	20	99	98	80700	4794	3823		21.11	2	1	100

Editable — Current selection set: 38 of 38

The PARCELTK FAT.

The PARCELTK FAT contains the Items from both the PARCELS theme and the CEDARBUF theme. Since you used an intersect overlay, all polygons in the theme are parcels within the buffer, so all future analysis will take place on the PARCELTK theme.

To calculate the cost of each portion of land in the PARCELTK theme, a new item must be added to the theme. The Calculate command can be used to multiply the square footage by the polygon area to come up with a total value per polygon.

```
Command: ADDITEM
Theme name (?/<parceltk>): ENTER
Item name (?): VALUE
Item width: 15
Item type?  (Character/Date/<Numeric>): ENTER
Item decimal places <0>: 2
Start item <last>: ENTER

Command: CALCULATE
Theme name (?/<parceltk>): ENTER
Target Item (?): VALUE
Area * SQFTVAL
```

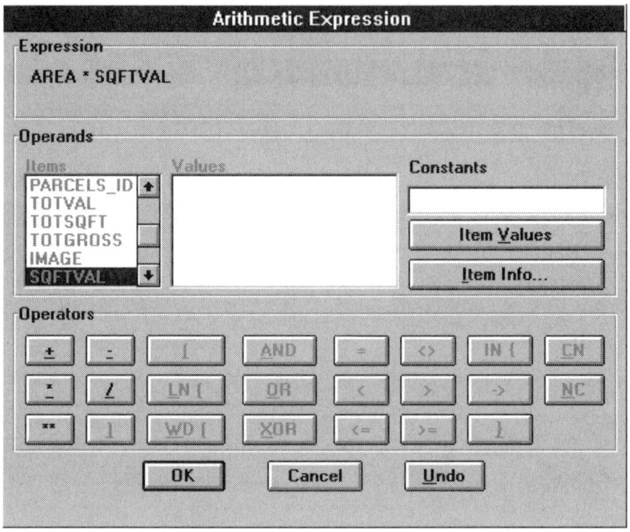

The Calculate expression.

Use the Wbrowse command now to view the PARCELTK FAT with the VALUE item calculated.

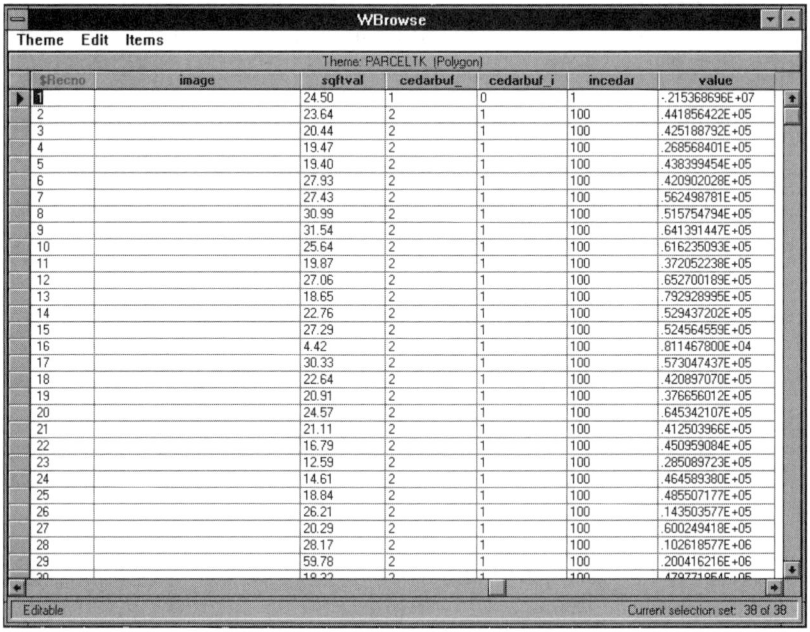

$Recno	image	sqftval	cedarbuf_	cedarbuf_i	incedar	value
1		24.50	1	0	1	-.215368696E+07
2		23.64	2	1	100	.441856422E+05
3		20.44	2	1	100	.425188792E+05
4		19.47	2	1	100	.268568401E+05
5		19.40	2	1	100	.438399454E+05
6		27.93	2	1	100	.420902028E+05
7		27.43	2	1	100	.562498781E+05
8		30.99	2	1	100	.515754794E+05
9		31.54	2	1	100	.641391447E+05
10		25.64	2	1	100	.616235093E+05
11		19.87	2	1	100	.372052238E+05
12		27.06	2	1	100	.652700189E+05
13		18.65	2	1	100	.792928995E+05
14		22.76	2	1	100	.529437202E+05
15		27.29	2	1	100	.524564559E+05
16		4.42	2	1	100	.811467800E+04
17		30.33	2	1	100	.573047437E+05
18		22.64	2	1	100	.420897070E+05
19		20.91	2	1	100	.376656012E+05
20		24.57	2	1	100	.645342107E+05
21		21.11	2	1	100	.412503966E+05
22		16.79	2	1	100	.450959084E+05
23		12.59	2	1	100	.285089723E+05
24		14.61	2	1	100	.464589380E+05
25		18.84	2	1	100	.485507177E+05
26		26.21	2	1	100	.143503577E+05
27		20.29	2	1	100	.600249418E+05
28		28.17	2	1	100	.102618577E+06
29		59.78	2	1	100	.200416216E+06
30		[illegible]	2	1	100	[illegible]

The PARCELTK FAT with value.

The next step in the process is to determine the sum of the VALUE item in the PARCELTK FAT. By calculating the sum of the VALUE item, the total estimated cost of the land taking can be established. Before the sum is calculated, the universe polygon must be eliminated from the selection set.

To eliminate the universe polygon, execute the Reselect command on the Query menu.

```
Command: RESELECT
Theme name (?/<parceltk>): ENTER
Reselect by?
  (Circle/ENtity/Expression/Polygon/Window): E
```

Reselect all polygons with area > 0.

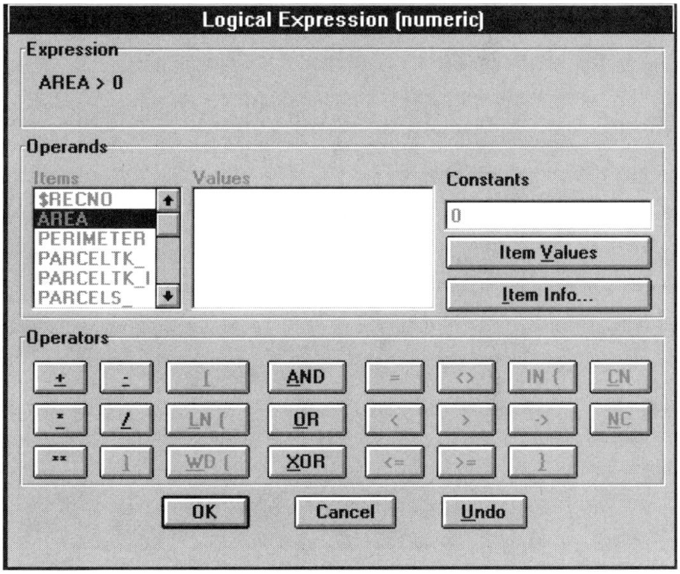


```
" AREA > 0" C:\INARC\PARCELTK polys : 37 of 38
  selected.
```

The Statistics command found on the Analysis menu can now be used to present the sum of the value item for the land taking.

```
Command: STATISTICS
Input Theme (?/<PARCELTK>): ENTER
Item (?): VALUE
```

```
Optional Case item (?/<none>): ENTER
Optional Output theme (?/<none>): ENTER
Total statistics for VALUE
COUNT          =          36
MINIMUM        =               .000000
MAXIMUM        =          48357.351200
SUM            =         483668.349910
SUM SQUARED    =      9353173626.100067
MEAN           =          13435.231942
STANDARD DEV.  =           9031.654115
```

To complete the first exercise, you need to generate a list of owners' names and addresses. This can be done by relating the LOTINFO theme, created in Chapter 7, to the PARCELTK theme. You previously used the Relate command. In this chapter you will use the Ddrelate command to join these two tables. Ddrelate simplifies the process of relating tables by using a dialog box interface. The dialog box, shown below, prompts for the From Theme and From Theme item, as well the To Theme and To Theme item.

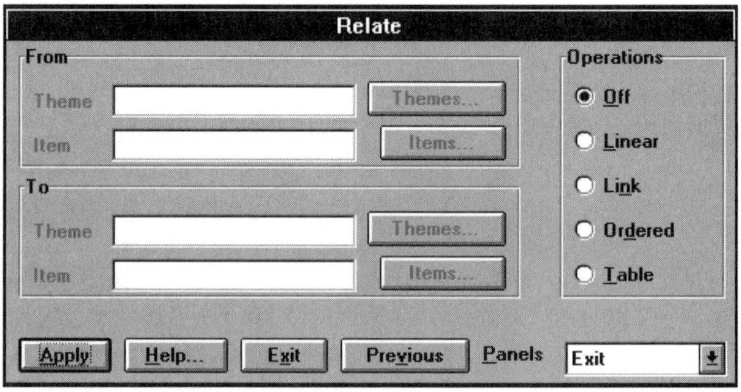

The Ddrelate dialog box.

To fill out the dialog box, start by selecting Linear as the Operation type. Selecting the operation type activates the From Theme button and edit box. Choose the Themes button to display the following dialog box.

The Ddrelate From Theme dialog box.

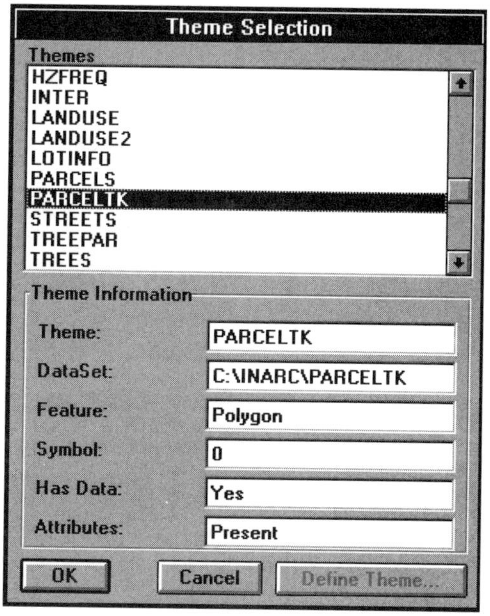

Select PARCELTK from the list of themes presented and press the OK button. From the From Theme area, select ITEM.

The Ddrelate From item dialog box.

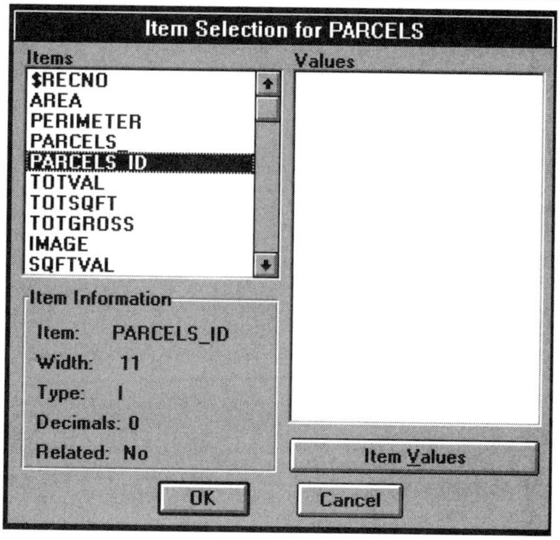

From the items list, select PARCELS_ID and press OK. With the From Theme information filled out, you must now specify the To Theme information.

From the To Theme area, select Themes.

The Ddrelate To Theme dialog box.

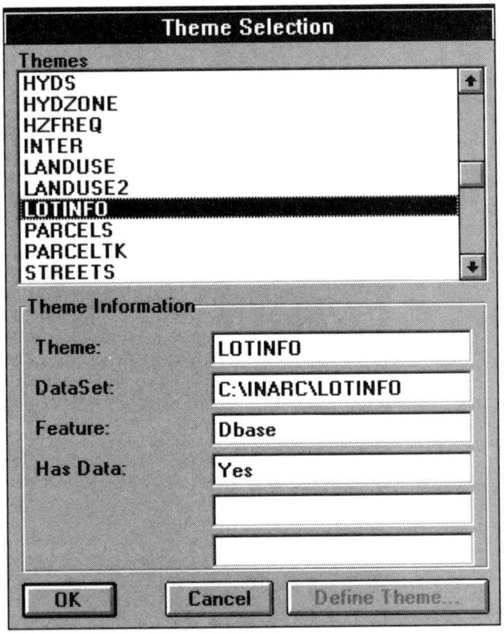

In the list presented, select LOTINFO and press OK. To complete the input, select ITEM from the To Theme area.

The Ddrelate To Item dialog box.

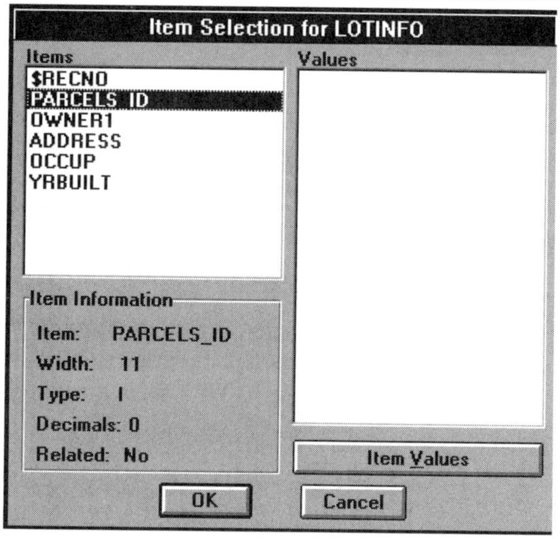

From this dialog box, select PARCELS_ID and press OK. The completed Ddrelate dialog box is shown below.

The completed Ddrelate dialog box.

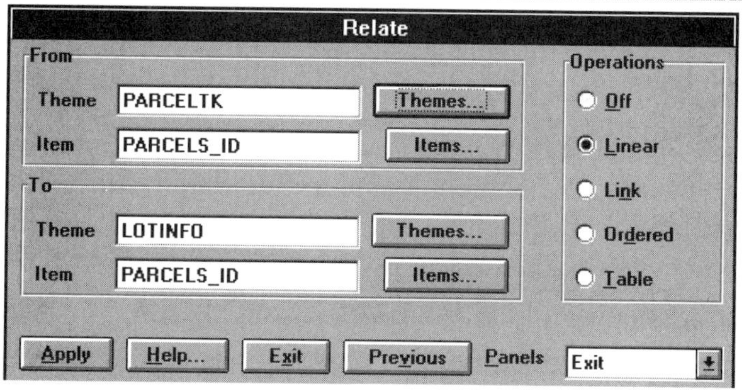

Once the dialog box is completed as shown, select Apply. ArcCAD makes the relate and returns you to the dialog box. Select Exit to dismiss the dialog box.

When relating tables as shown above, the From Item and To Item names do not need to be the same.

The related table now shows each parcel affected by the road widening, along with the owners' names and addresses. This table could now be used as the input to a report writing program to create a form letter to each of the names in the database.

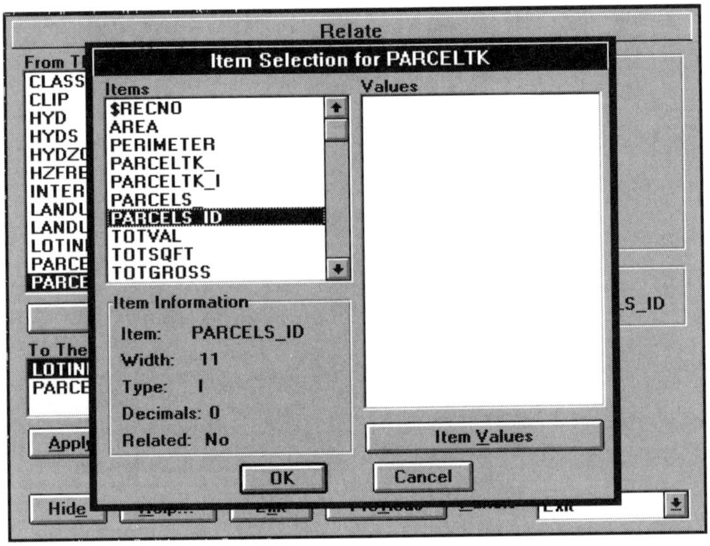

The Items subdialog box of Ddrelate.

To complete the relation, select LOTINFO from the theme list and select Apply. To exit the dialog box, select OK.

The related table now shows each parcel affected by the road widening, along with the owners' name and address. This table could now be used as input to a report writing program to create a form letter to each of the names in the database.

You have now competed the first exercise in analyzing spatial data. You have used the Buffer and Overlay commands to analyze a set of spatial data and now have a series of tables that can be further analyzed using thematic mapping tools or with external database and spreadsheet applications. At this point, ArcView can also be used to create reports and presentations. In the next section you will use ArcCAD's spatial analysis tools to create point in polygon overlays and to generate frequency reports.

Comparing Fire Hydrant Locations

The next spatial analysis example is to compare the number of fire hydrants in each zone. This type of analysis uses a point in polygon intersect, the result of which can be used as the input to the Frequency command.

First, use the AutoCAD Erase command to clear the screen.

Next, set the ZONING layer current and execute the Polys command found on the Display pulldown under Polygons to display the ZONING polygons.

```
Command: LA
LAYER ?/Make/Set/New/ON/OFF/Color/Ltype/Freeze/Thaw/
  LOck/Unlock: M
New current layer <BUFFER>: ZONING
?/Make/Set/New/ON/OFF/Color/Ltype/Freeze/Thaw/LOck/
  Unlock: ENTER

Command: POLYS
Theme name (?/<PARCELTK>): ZONING
Line symbol number (0-100) (?/<0>): 6
```

With the ZONING polygons in the drawing, the next step will be to extract the fire hydrants from the UTILS theme so that they may be intersected with the ZONING polygons.

The extraction process consists of two parts. The first step is to run a query that selects all of the hydrants from the UTILS theme.

```
Command: RESELECT
Theme name (?/<ZONING>): UTILS
Reselect by?
  (Circle/ENtity/Expression/Polygon/Window): E
```

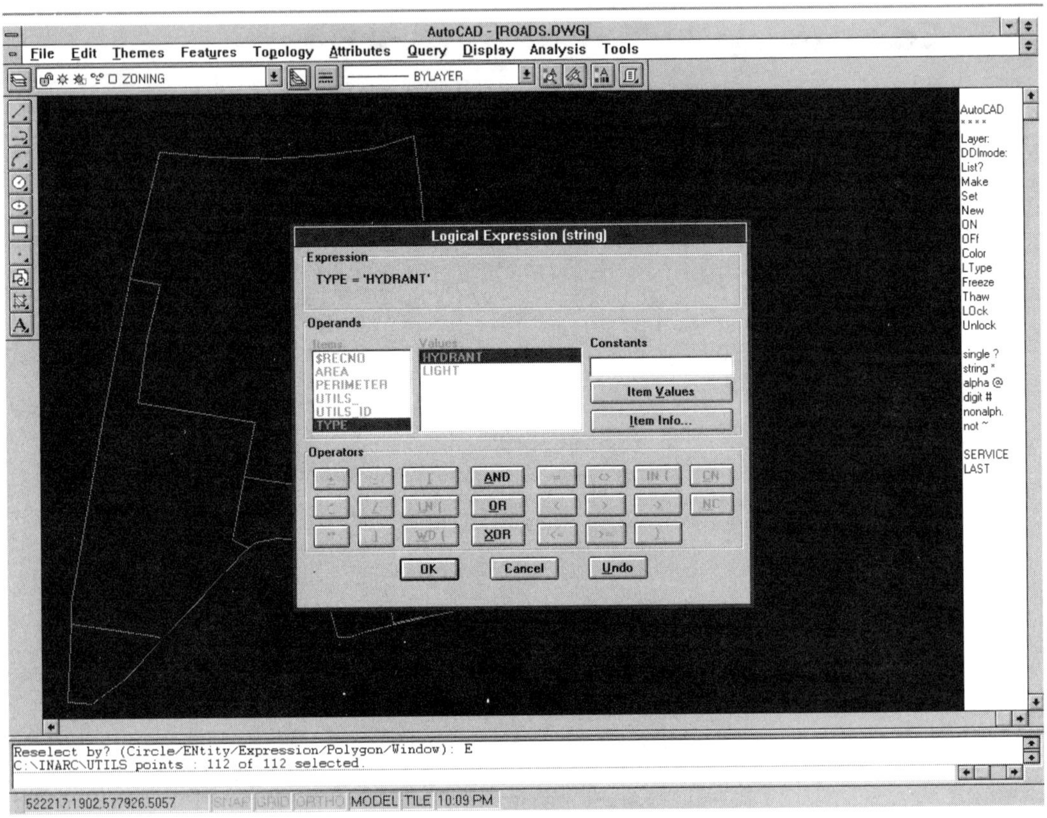

Reselect all hydrants from UTILS.

```
C:\INARC\UTILS points : 112 of 112 selected.
```

In the Expression dialog box, enter `TYPE = HYDRANT` as shown above.

> **NOTE:** *Selecting the Item Values button will display a list of unique item values in the Values Selection box. Pick Item values and then double-click on HYDRANT in the list.*

```
" TYPE = 'HYDRANT'" C:\INARC\UTILS points : 36 of 112
  selected.
```

With the HYDRANTS selected, execute the Extract command found on the Analysis menu.

```
Command: EXTRACT
Input theme (?/<UTILS>): ENTER
Output theme (?): HYDS
```

```
Theme HYDS is not defined.
```

Select OK from the Define Theme dialog box.

```
Defining HYDS...
Extracting features from UTILS to create HYDS.
36 of 112 records selected.
Extracting points...
Number of Points in Coverage...     112
Number of Points Reselected....      36
Creating Point attribute table.
```

The extract command has created a new theme name HYDS and written the selected features in UTILS to it.

The ZONING and HYDS themes can now be intersected to determine which zone each of the hydrants is in. Execute the ERSI_INTERSECT command and follow the input below.

```
Command: ESRI_INTERSECT
Input theme (?/<HYDS>): ENTER
Intersect theme (?): ZONING
Output theme (?): HYDZONE
Theme HYDZONE is not defined.
```

Select OK from the Define Theme dialog box.

```
Defining HYDZONE...
Fuzzy tolerance <compute>: ENTER
Intersecting HYDS with ZONING to create HYDZONE.
Overlaying points...
Creating Point attribute table.
Item "AREA" duplicated, Join File version dropped
Item "PERIMETER" duplicated, Join File version dropped
Item "HYDS_" duplicated, Join File version dropped
Item "AREA" duplicated, Join File version dropped
Item "PERIMETER" duplicated, Join File version dropped
Item "ZONING_" duplicated, Join File version dropped
```

The intersect process has added the zone name to each of the hydrants in the HYDS theme. To view the FAT for HYDZONE, execute the Wbrowse command.

WBrowse

Theme Edit Items

Theme: HYDZONE (Point)

$Recno	area	perimeter	hydzone_	hydzone_id	hyds_	hyds_id	type	zoning_	zoning_id	zone
1	.0000000E+00	.0000000E+00	1	1	3	50	HYDRANT	2	3	RB
2	.0000000E+00	.0000000E+00	2	2	4	51	HYDRANT	2	3	RB
3	.0000000E+00	.0000000E+00	3	3	5	52	HYDRANT	2	3	RB
4	.0000000E+00	.0000000E+00	4	4	6	53	HYDRANT	2	3	RB
5	.0000000E+00	.0000000E+00	5	5	25	101	HYDRANT	2	3	RB
6	.0000000E+00	.0000000E+00	6	6	1	5	HYDRANT	2	3	RB
7	.0000000E+00	.0000000E+00	7	7	7	54	HYDRANT	3	2	RC
8	.0000000E+00	.0000000E+00	8	8	26	102	HYDRANT	4	7	C&A
9	.0000000E+00	.0000000E+00	9	9	8	55	HYDRANT	3	2	RC
10	.0000000E+00	.0000000E+00	10	10	9	56	HYDRANT	2	3	RB
11	.0000000E+00	.0000000E+00	11	11	10	57	HYDRANT	2	3	RB
12	.0000000E+00	.0000000E+00	12	12	11	58	HYDRANT	2	3	RB
13	.0000000E+00	.0000000E+00	13	13	12	59	HYDRANT	2	3	RB
14	.0000000E+00	.0000000E+00	14	14	16	63	HYDRANT	3	2	RC
15	.0000000E+00	.0000000E+00	15	15	32	108	HYDRANT	2	3	RB
16	.0000000E+00	.0000000E+00	16	16	15	62	HYDRANT	3	2	RC
17	.0000000E+00	.0000000E+00	17	17	27	103	HYDRANT	4	7	C&A
18	.0000000E+00	.0000000E+00	18	18	14	61	HYDRANT	3	2	RC
19	.0000000E+00	.0000000E+00	19	19	13	60	HYDRANT	2	3	RB
20	.0000000E+00	.0000000E+00	20	20	17	64	HYDRANT	3	2	RC
21	.0000000E+00	.0000000E+00	21	21	18	65	HYDRANT	3	2	RC
22	.0000000E+00	.0000000E+00	22	22	35	111	HYDRANT	5	6	CBD
23	.0000000E+00	.0000000E+00	23	23	36	112	HYDRANT	5	6	CBD

Editable

Current selection set: 36 of 36

The HYDZONE FAT.

To calculate the number of hydrants in each zone, use the Frequency command found on the Analysis menu.

```
Command: FREQUENCY
Input theme (?/<HYDZONE>): ENTER
Output theme (?): HZFREQ
Theme HZFREQ is not defined.
```

Select OK from the Define Theme dialog box.

```
Defining HZFREQ...
Case item (?/<CASE_>): ENTER
Frequency Items (?/<END>): ZONE
Frequency Items (?/<END>): ENTER
Summary Items (?/<END>): ENTER
```

The Frequency command creates a new record theme that lists the number of times a fire hydrant shows up in each zone. The CASE_ ITEM is added to the input theme and contains the record number of the output theme to which the input theme record was classified. You will use this CASE_ ITEM later in this chapter to relate the input theme to the output theme.

To view the results of the Frequency command, execute the Wbrowse command on the HZFREQ theme.

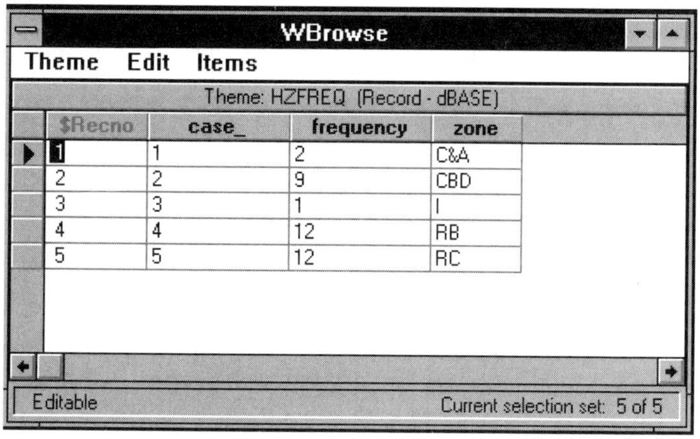

The HZFREQ record database.

Further analysis can be done for the HZFREQ theme using any application that can read dBASE III format files. For example, ArcView could be used to create a chart of the number of hydrants in each zone.

Managing Utility Easements

In this section of the chapter, you will learn to do the following:

- ❐ Buffering of trees based on an item in the database.
- ❐ Buffer of the utility lines.
- ❐ Union of the tree buffer with the utility buffer.
- ❐ Reselection and extraction of trees in the buffer.
- ❐ Intersect of the extraction with parcels for owner notification.
- ❐ Relation of new theme with Lotinfo.

The last example in spatial analysis will be to determine the location of utility easements. The easements you are interested in are the ones located on private land. You will begin this project by identifying the easement locations. The project will be completed in the next chapter by creating a thematic map showing the easement locations on a map.

The analysis will include the overlay of the utility lines on the parcels. Once the overlay has been created, a buffer representing the easement width will be created for the utility lines.

To create the overlay of utility lines and parcels, execute the ESRI_INTERSECT command found on the Analysis menu.

Review

In this chapter you have begun to query and analyze data with ArcCAD. Commands that limit and include specific amounts of data are important for you to be able to work effectively with ArcCAD. With the Buffer, IDENTITY, ESRI_INTERSECT, and ESRI_UNION commands, you can access the information necessary for your analysis.

Part III.

Thematic Mapping

Chapter 10

Queries

Building Selection Sets

In this chapter you will learn to select ArcCAD features based on both attributes and location queries. You will learn the components and method of creating logical expressions.

You can reduce the amount of topological analysis to be done by using the correct combination of location and attribute queries.

Setup

To begin this chapter,

> Load AutoCAD and open the INARC10 drawing file. This drawing file has the PARCELS, LANDUSE, STREETS, ULINE, and TREES themes defined. A layer for each theme has been created and the extents of the drawing have been set.

Querying and Selecting Features in Themes

You will begin this chapter by investigating how ArcCAD queries or selects features within a theme. You have already seen some of these commands in previous chapters. For example, the Reselect command was used in Chapter 7 to extract Cedar Street from the STREETS theme.

All of the commands that will be discussed in this section can be found on the Query pulldown menu shown in the following figure.

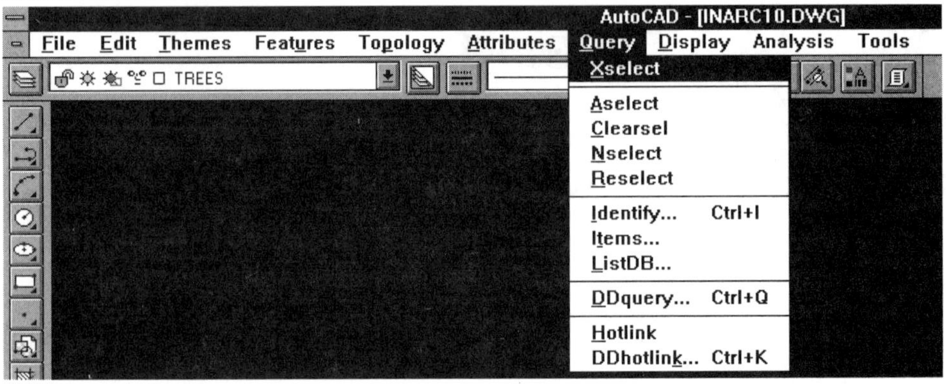

The Query pulldown menu.

The following are the four query commands used to build and manipulate selection sets in ArcCAD:

- **Reselect:** The Reselect command selects a subset of the currently selected records based on feature attributes or location. The options available for the Reselect command are Circle, Polygon, Window, Entities, and Expression. The first four options in the list are query by location commands. The fifth is the query by Attributes command.
- **Aselect:** Aselect adds to the current selection set by querying the unselected records. Aselect also allows you to select features based on either attributes or location, and will supply the same options as the Reselect command.
- **Nselect:** The Nselect command switches the currently selected records with the unselected records.
- **Clearsel:** The Clearsel command clears the current selection set. Clearsel is a global command and clears every theme's current selection sets.

Selecting Features by Expression

An example of a query by expression would be to select all of the street centerlines with a class of 5.

```
Command: RESELECT
Theme name (?/<THEME>): Streets
Reselect by?
  (Circle/ENtity/Expression/Polygon/Window): E
```

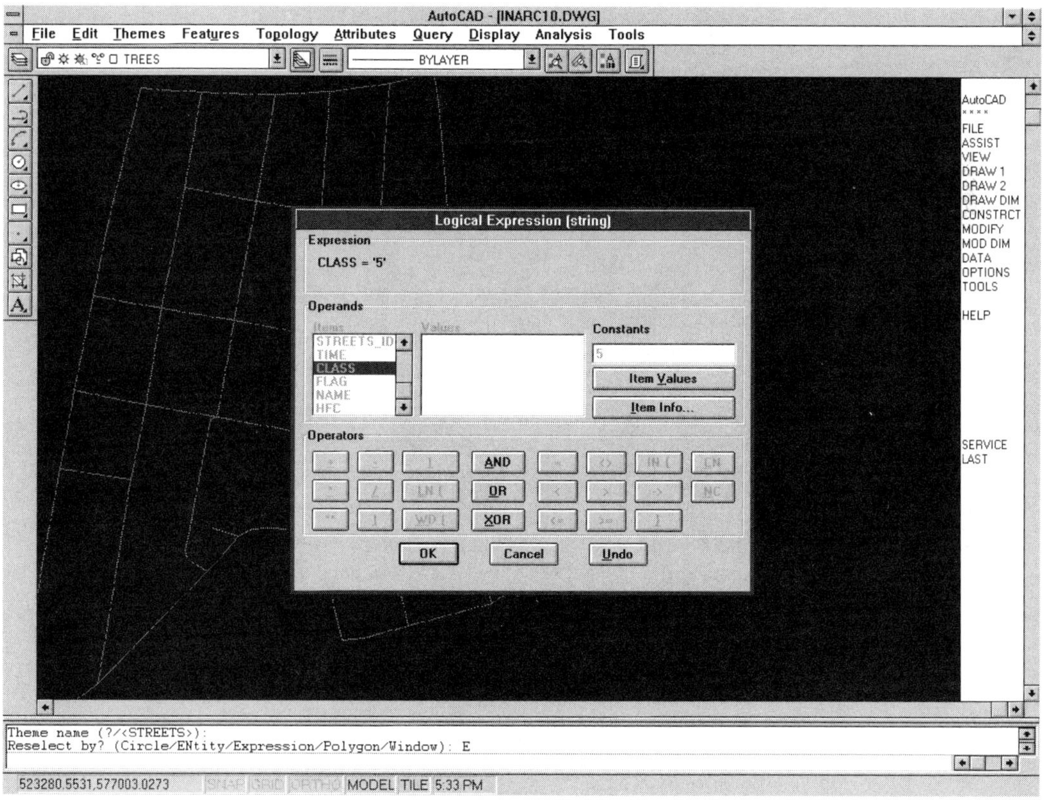

The Reselect Expression dialog box.

Double-click on Class in the operands, = in the operators; type 5 in the Constants edit box.

```
" CLASS = '5'" C:\INARC\STREETS arcs : 44 of 70
  selected.
```

Reselect has built a selection set of 44 out of a total of 70 centerlines.

Once a selection set is created, it remains the current selection for all subsequent commands.

The Aselect command can then be used to add unselected features to the current selection set.

```
Command: ASELECT
Theme name (?/<STREETS>):
Add selection
  by?(Circle/ENtity/Expression/Polygon/Window):E
```

The Aselect Expression dialog box.

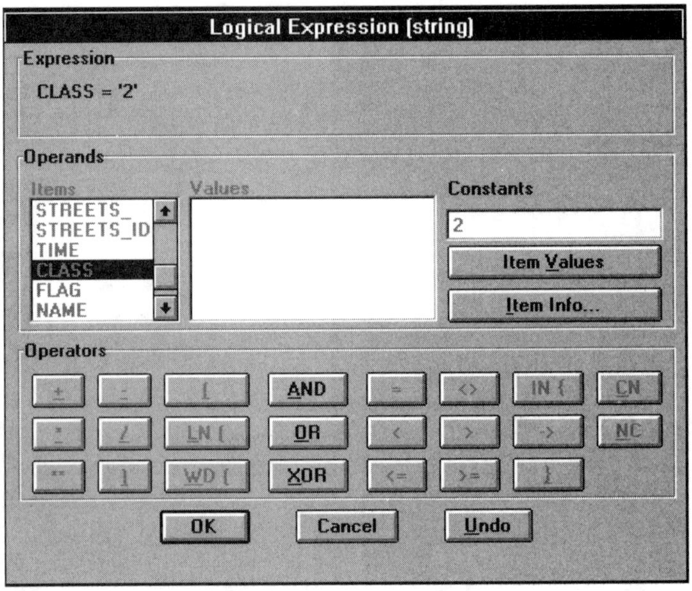


```
" CLASS = '2'" C:\INARC\STREETS arcs : 12 of 70
  selected.
```

The Aselect command has added 12 records to the current selection set.

To view the records, execute the Wbrowse command. At the Theme Name prompt, enter STREETS.

```
Command: WBrowse
Theme name (?/<STREETS>): ENTER
56 of 70 records selected.
```

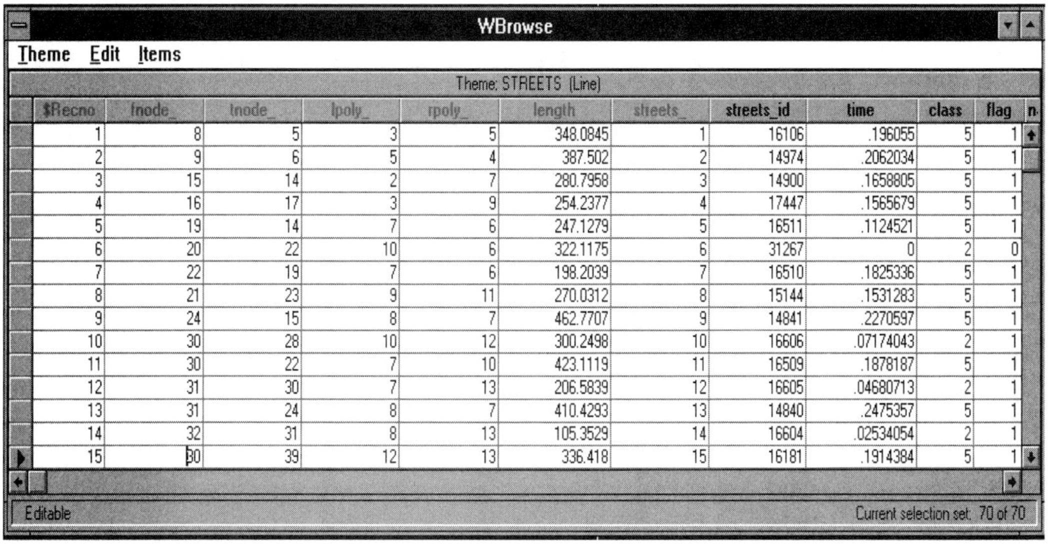

WBrowse

Theme Edit Items

Theme: STREETS (Line)

$Recno	fnode_	tnode_	lpoly_	rpoly_	length	streets_	streets_id	time	class	flag	n
1	8	5	3	5	348.0845	1	16106	.196055	5	1	
2	9	6	5	4	387.502	2	14974	.2062034	5	1	
3	15	14	2	7	280.7958	3	14900	.1658805	5	1	
4	16	17	3	9	254.2377	4	17447	.1565679	5	1	
5	19	14	7	6	247.1279	5	16511	.1124521	5	1	
6	20	22	10	6	322.1175	6	31267	0	2	0	
7	22	19	7	6	198.2039	7	16510	.1825336	5	1	
8	21	23	9	11	270.0312	8	15144	.1531283	5	1	
9	24	15	8	7	462.7707	9	14841	.2270597	5	1	
10	30	28	10	12	300.2498	10	16606	.07174043	2	1	
11	30	22	7	10	423.1119	11	16509	.1878187	5	1	
12	31	30	7	13	206.5839	12	16605	.04680713	2	1	
13	31	24	8	7	410.4293	13	14840	.2475357	5	1	
14	32	31	8	13	105.3529	14	16604	.02534054	2	1	
15	30	39	12	13	336.418	15	16181	.1914384	5	1	

Editable

Current selection set: 70 of 70

The STREETS FAT.

The WBrowse command displays only the selected records. Pressing the END key will move one screen to the right. Notice that all of the class values are 5 or 2. Press ESC to exit.

To switch the current selection set with the unselected records, execute the Nselect command.

```
Command: NSELECT
Theme name (?/<STREETS>): ENTER
C:\INARC\STREETS arcs : 14 of 70 selected.
```

To view the new selection set, execute the Listdb command again.

```
Command: LISTDB
Theme name (?/<STREETS>): ENTER
14 of 70 records selected.
```

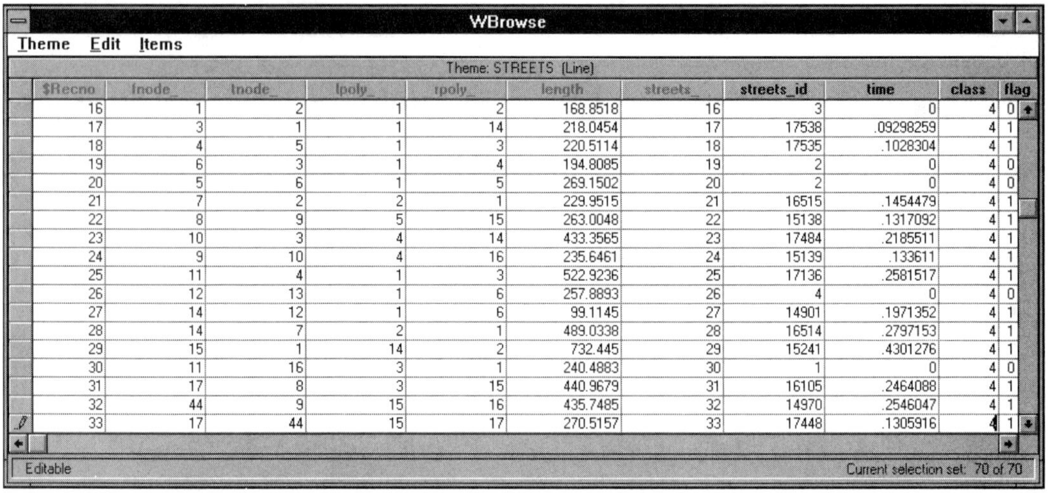

$Recno	lnode_	tnode_	lpoly_	rpoly_	length	streets_	streets_id	time	class	flag
16	1	2	1	2	168.8518	16	3	0	4	0
17	3	1	1	14	218.0454	17	17538	.09298259	4	1
18	4	5	1	3	220.5114	18	17535	.1028304	4	1
19	6	3	1	4	194.8085	19	2	0	4	0
20	5	6	1	5	269.1502	20	2	0	4	0
21	7	2	2	1	229.9515	21	16515	.1454479	4	1
22	8	9	5	15	263.0048	22	15138	.1317092	4	1
23	10	3	4	14	433.3565	23	17484	.2185511	4	1
24	9	10	4	16	235.6461	24	15139	.133611	4	1
25	11	4	1	3	522.9236	25	17136	.2581517	4	1
26	12	13	1	6	257.8893	26	4	0	4	0
27	14	12	1	6	99.1145	27	14901	.1971352	4	1
28	14	7	2	1	489.0338	28	16514	.2797153	4	1
29	15	1	14	2	732.445	29	15241	.4301276	4	1
30	11	16	3	1	240.4883	30	1	0	4	0
31	17	8	3	15	440.9679	31	16105	.2464088	4	1
32	44	9	15	16	435.7485	32	14970	.2546047	4	1
33	17	44	15	17	270.5157	33	17448	.1305916	4	1

The STREETS FAT.

In the new selection set, only the records that are not class 2 or 5 are selected.

Attribute Queries

When using the Reselect or Aselect command to build a query by expression, the command displays an Expression Builder dialog box. This dialog box makes the building of complex logical expressions as simple as pointing and clicking.

The syntax of a logical expression is as follows:

```
Item name, Operator, Item value
```

In the previous Reselect example you created the following logical expression:

```
Item name = "Class", Operator = "=", Item value = "5"
```

When creating attribute queries, a logical operand may optionally be used to connect multiple queries. The logical operands available are And, Or, and Xor. To join the two previous Reselect and Aselect examples into one query, the following expression could be used:

```
Class = 5 OR Class = 2.
```

To try this query, execute the Clearsel command found on the Query menu. Next, execute the Reselect command and follow the input below.

```
Command: RESELECT
Theme name (?/<STREETS>): ENTER
Reselect by?
  (Circle/ENtity/Expression/Polygon/Window): E

" CLASS = '5' OR CLASS = '2'" C:\INARC\STREETS arcs :
  56 of 70 selected.
```

Location Queries

The Reselect and Aselect options for creating location queries are Window, Polygon, and Circle. These commands allow features to be selected based on a defined area or a distance from a selected point. An example of a location query would be to select all of the fire hydrants within 300 feet of a certain parcel or all street lights in a selected city block.

To try this type of query, you will first need to display the features for the BLDGS and UTILS in the drawing.

Execute the Polys command from the Display pulldown menu under Polygons.

```
Command: POLYS
Theme name (?/<STREETS>): PARCELS
Line symbol number (0-100) (?/<0>): 4
```

Next, select the Pointmrk command from the Display menu under Points.

```
Command: POINTMRK
Theme name (?/<BLDGSCLP>): UTILS
Display using? (Properties/Item/<Symbol>): ENTER
Marker symbol number (0-100) (?/<0>): 22
```

The first of the location query commands you will use is the Circle option of the Reselect command. This command allows a spatial query to be made by selecting a point and specifying a search radius. All features of the input theme within the search radius will be selected. To demonstrate a Circle query, you will select all hydrants within 300 feet of the parcel owned by Robert T. Leclair.

The first step in creating this query is to relate the PARCELS theme to the LOTINFO record theme. This allows you to find the lot owned by LeClair.

```
Command: RELATE
Relate? (OFF/<ON>): ON
From theme (?/<UTILS>): PARCELS
To theme (?): LOTINFO
Relate item (?): PARCELS_ID
Relate method? (Link/Ordered/Table/<LInear>): ENTER
```

Once the relate has been completed, execute the Reselect command on the Query menu to do an attribute query on "Leclair Robert T."

```
Command: RESELECT
Theme name (?/<PARCELS>): ENTER
Reselect by?
  (Circle/ENtity/Expression/Polygon/Window): E
C:\INARC\PARCELS polys : 351 of 351 selected.
```

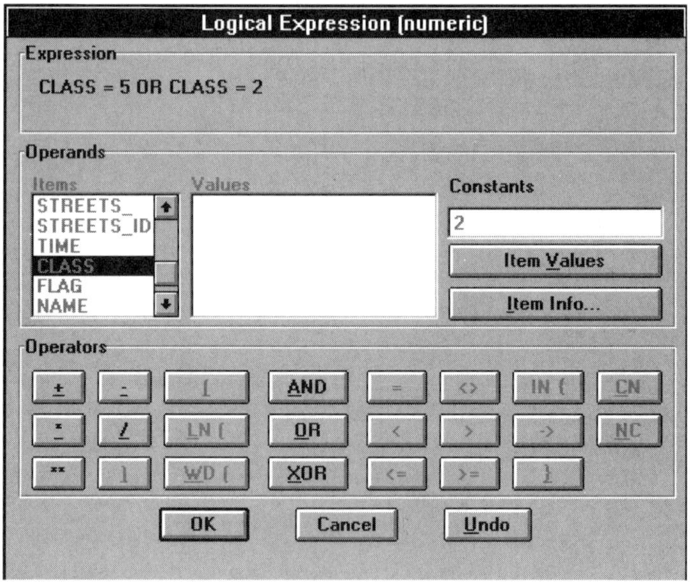

The Reselect by Expression dialog box.

Double-click on #Owner1, single-click on =. Select the ItemValues button. From the Values list, double-click on LeClair Robert T. and press OK.

```
" #OWNER1 = 'LECLAIR  ROBERT T.'"
C:\INARC\PARCELS polys : 1 of 351 selected.
```

To display this parcel on screen use the Polyshd command on the Display pulldown menu under Polygons.

```
Command: POLYSHD
Theme name (?/<PARCELS>): ENTER
Display using? (Properties/Item/<Symbol>): ENTER
Shade symbol number (0-100) (?/<4>): ENTER
```

With the display complete, you are ready to create the circle query. Execute the Reselect command from the Query menu.

```
Command: RESELECT
Theme name (?/<PARCELS>): UTILS
Reselect by?
  (Circle/ENtity/Expression/Polygon/Window): C
Center point:
```

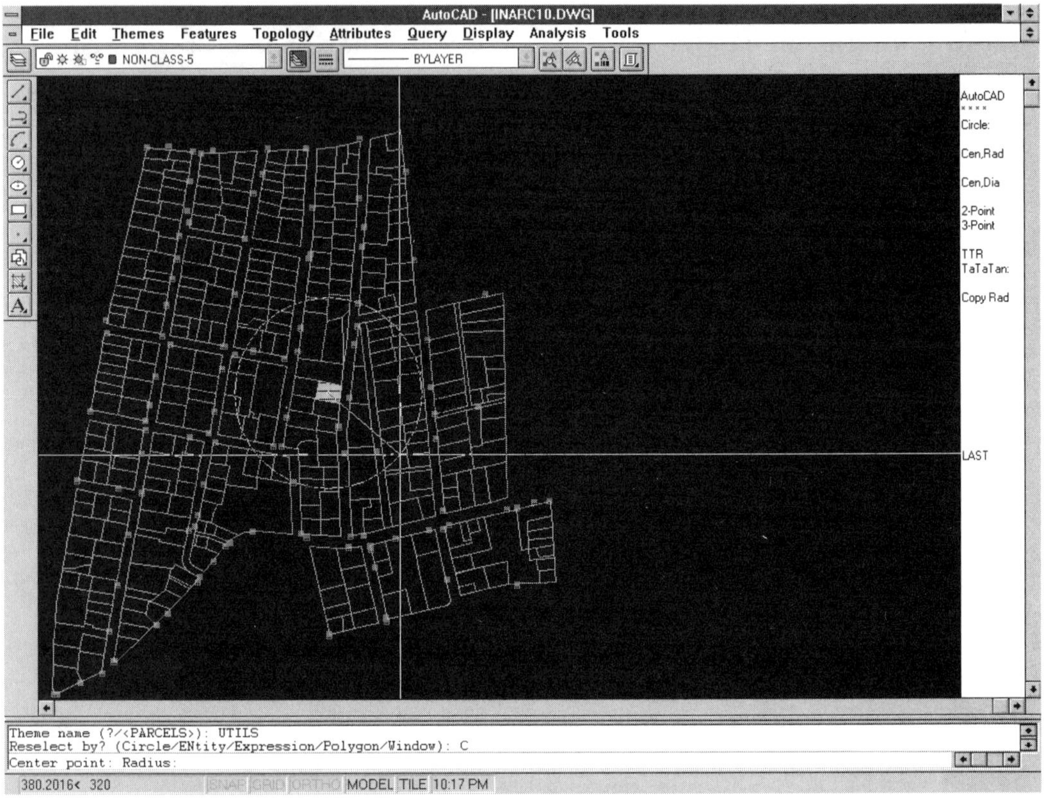

Selecting the centerpoint of the highlighted parcel.

```
Radius: 300
Within/<Crossing>: ENTER
C:\INARC\UTILS points : 10 of 112 selected.
```

The Within/<Crossing> prompt determines whether to select only features that are completely within the radius or features that are completely and partially within the radius.

ArcCAD has selected 10 out of 112 features from the UTILS theme. However, the features selected are both hydrants and lights.

To reduce the selection set to just the hydrants, execute the Reselect command and query for TYPE = HYDRANT as shown below.

```
Command: RESELECT
Theme name (?/<UTILS>): ENTER
```

```
Reselect by?
  (Circle/ENtity/Expression/Polygon/Window): E
```

The Reselect dialog box.

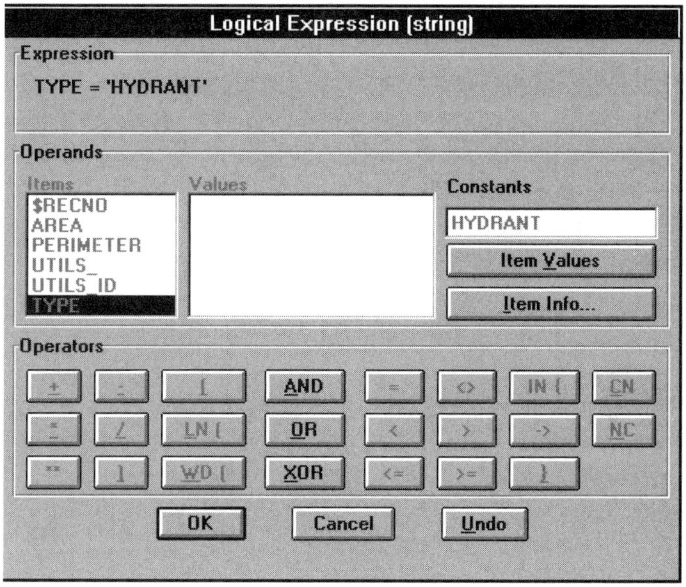


```
" TYPE = 'HYDRANT'" C:\INARC\UTILS points : 3 of 112
  selected.
```

You have completed the process of selecting all hydrants within 300 feet of a selected parcel.

The last two of the location query command options allow features to be selected using either a user-defined window or polygon. These two options are identical to the AutoCAD object selection method. Both provide options to select features inside or crossing the window or polygon.

The last of the Reselect query command options is the Entity option. This option allows you to select features based on an entity selection set. All of the AutoCAD selection methods are supported.

Review

The query chapter has introduced you to the commands needed to create selection sets. Selection sets have been built based on attribute queries as well as location queries. You have seen how the amount of topological analysis can be reduced by using the right combination of location and attribute queries and you have learned the structure of a logical expression. Remember, once a selection set has been created, it remains the current selection set until it is either cleared using the Clearsel command or a new query adds to or removes features from it.

Chapter

Display

Creating Drawings from Themes

In this chapter you will learn how ArcCAD displays entities in AutoCAD. Concepts covered include how ArcCAD handles symbology and how AutoCAD links entities to the ArcCAD spatial and tabular data sets. You will use commands in this chapter that display lines, points, and polygon shading.

Setup

To begin this chapter,

> Load AutoCAD and open the INARC10 drawing file. This drawing file has the PARCELS, LANDUSE, STREETS, ULINE, and TREES themes defined, a layer for each theme has been created, and the extents of the drawing have been set. If any information exists in the drawing from the previous chapter, erase it now.
>
> To finish the setup, execute the Polys command to display the Parcels in the drawing.

```
Command: POLYS
Theme name (?/<THEME>): PARCELS
Line symbol number (0-100) (?/<0>): 4
```

Symbol Sets

Symbol sets are databases of information that control the way ArcCAD displays features in AutoCAD. These databases contain the list of AutoCAD properties that ArcCAD will use to display a particular feature. For example, when displaying Arclines for a line theme with symbol number 22, ArcCAD looks in the line symbol database and retrieves the color and linetype definition for symbol 22. It then uses this information to display the lines.

ArcCAD defines four different sets of symbols for use in displaying features. The four sets are

- Line Symbols
- Marker Symbols
- Shade Symbols
- Text Symbols

The various symbol sets may be displayed by using the LinIndex, MrkIndex, ShdIndex, and TxtIndex commands, and the definition of each of the symbol sets may be viewed using the Syminfo command.

Each of these commands can be found on the Display pulldown under the Symbols pullout.

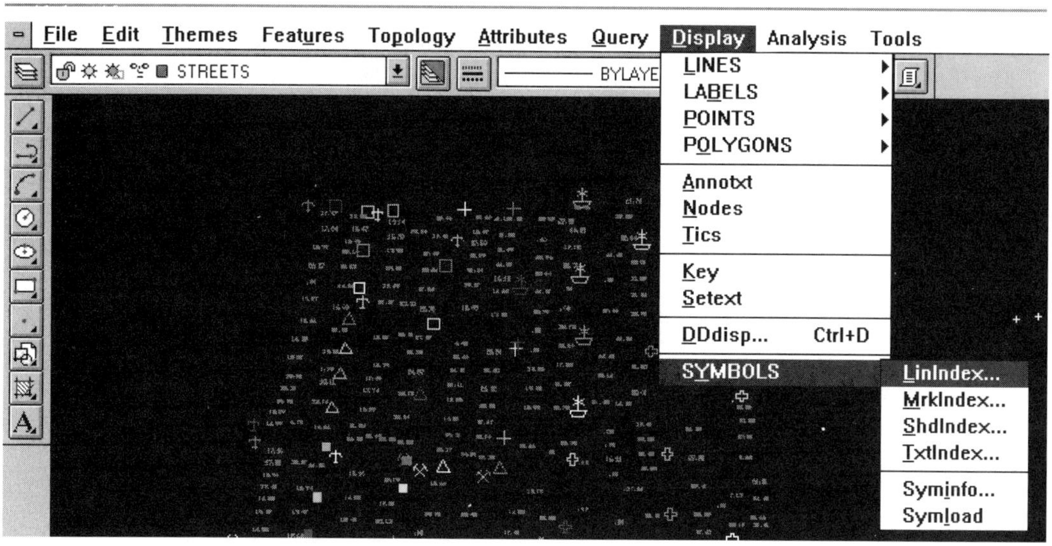

The Symbols menu.

The symbol sets may also be displayed from within any command that prompts for a symbol number by entering a question mark at the symbol number prompt.

To view the line symbols now, execute the Arclines command for the STREETS theme.

```
Command: ARCLINES
Theme name (?/<THEME>): STREETS
Line symbol number (0-100) (?/<0>): ?
```

The Syminfo dialog box for Line symbols.

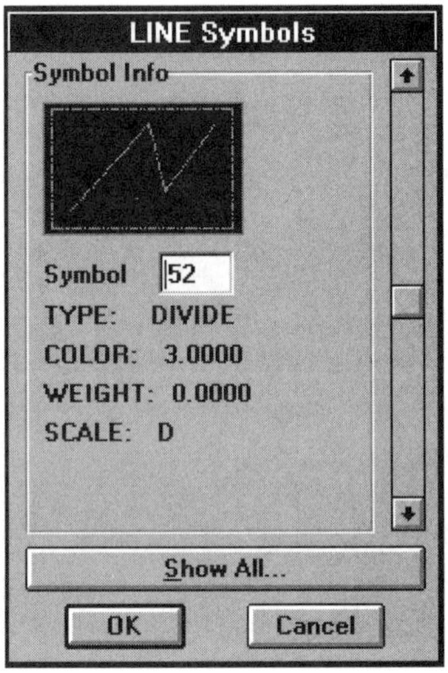

Select Show All from the Line Symbol dialog box.

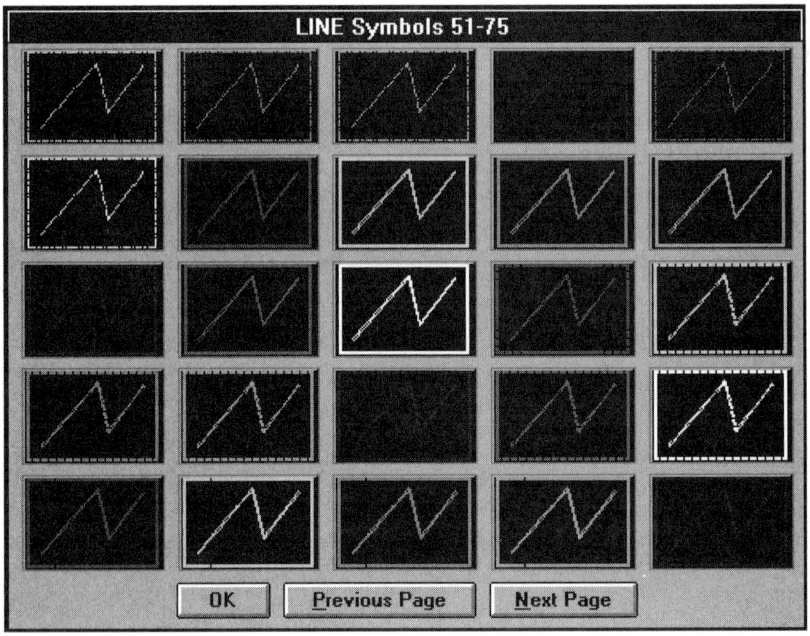

The SymIndex dialog box for Line symbols.

Each of the symbols shown is referenced by a number from 1 to 100. Double-click on the third symbol on the first row of the Lines Symbols 1-25 dialog box. This returns you to the Line Symbol dialog box and displays an image of the selected symbol and the symbol number 3. To accept this symbol select OK.

ArcCAD draws the streets into the drawing on the current layer, with the line symbol selected.

Symbol Number 0

Each symbol number is assigned a color, along with either a linetype, hatch pattern, block name, or text font. This is true for all symbols, except symbol number 0. Symbol 0 in ArcCAD means Bylayer. When displaying features in ArcCAD using symbol 0, the features will take on the properties of the current layer.

To see this, erase all entities from the drawing using the AutoCAD Erase command. Set the STREETS layer current, change its color to red, and its linetype to phantom.

```
Command: LA
LAYER ?/Make/Set/New/ON/OFF/Color/Ltype/Freeze/Thaw/
  LOck/Unlock: S
New current layer <0>: STREETS
?/Make/Set/New/ON/OFF/Color/Ltype/Freeze/Thaw/LOck/
  Unlock: C
Color: 1
Layer name(s) for color 1 (red) <STREETS>: ENTER
?/Make/Set/New/ON/OFF/Color/Ltype/Freeze/Thaw/LOck/
  Unlock: LT
Linetype (or ?) <CONTINUOUS>: PHANTOM
Layer name(s) for linetype PHANTOM <STREETS>: ENTER
?/Make/Set/New/ON/OFF/Color/Ltype/Freeze/Thaw/LOck/
  Unlock: ENTER
```

Execute the Arclines command to display the STREET centerlines.

```
Command: ARCLINES
Theme name (?/<STREETS>): ENTER
Display using? (Properties/Item/<Symbol>): ENTER
Line symbol number (0-100) (?/<22>): 0
```

The drawing should look like the following figure.

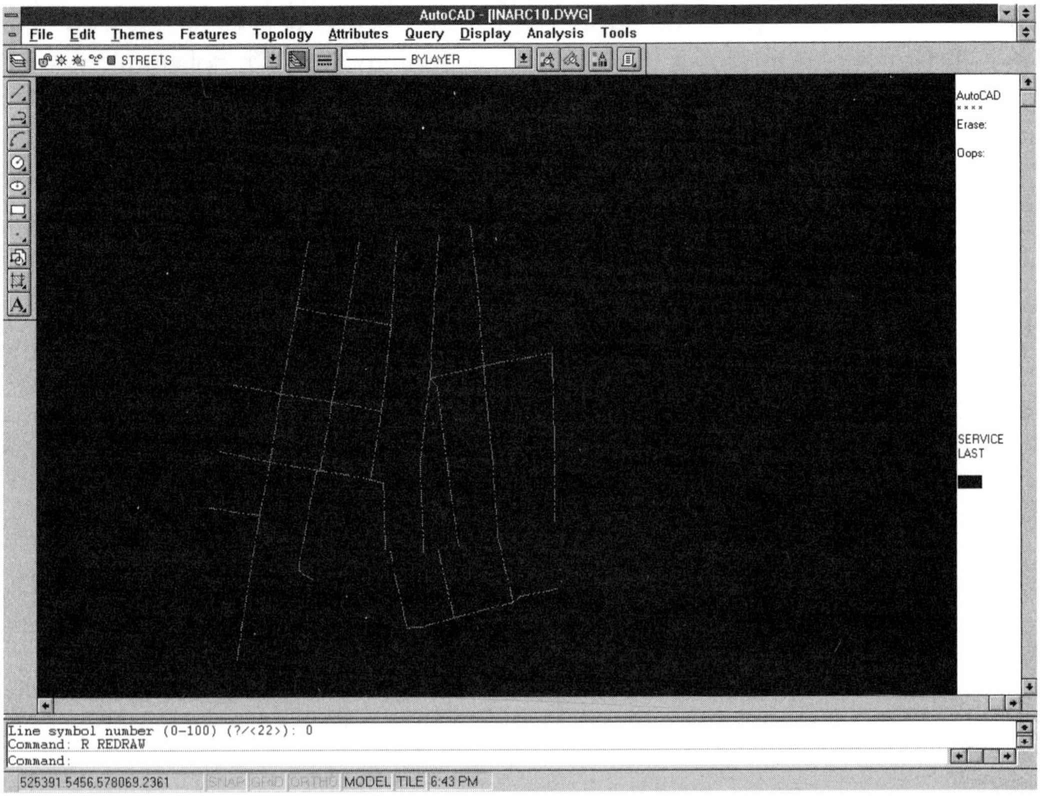

The drawing with centerlines displayed on the STREETS layer.

You should notice that the centerlines are displayed on the current layer, STREETS, using the color and linetype defined for that layer.

Default Symbols

The Defthm dialog box allows the input of a default symbol number. This symbol number will be presented as the default each time a theme is displayed. For example, when the PARCELS theme was defined, the default symbol was set to symbol 40. When displaying shade for the parcels theme, the default of 40 will be presented.

To demonstrate this, execute the Polyshd command found on the Display pulldown menu under Polygons.

```
Command: POLYSHD
Theme name (?/<STREETS>): PARCELS
Display using? (Properties/Item/<Symbol>): ENTER
Shade symbol number (0-100) (?/<40>): ENTER
```

Notice that the default Line Symbol number is set to 40. These settings shade the parcels with a blue, diagonal hatch.

Use the AutoCAD Undo command now to remove the shade from the drawing.

ArcCAD's ability to define symbols helps and hinders your work. Care should be taken when displaying features using a symbol number other than 0. ArcCAD's symbol assignments make creating complete drawings on layer 0 too easy. Working on these drawings later on can be cumbersome.

✔ **TIP:** *Making a habit of using symbol number 0 forces the user to create the correct layers and symbology and makes working with the database much easier.*

Entity Feature Links

Entity feature links are used by ArcCAD to connect AutoCAD entities to records in the spatial database. These links are used when identifying and modifying FAT records and in the graphical editing of features. Only entities with entity feature links can be edited graphically.

When creating an entity feature link, ArcCAD writes common information to three separate locations. The common item used to link the entities to the theme is the User-ID. This ID is written to the spatial database, the FAT, and the extended entity data of the AutoCAD entity. The Link Mode setting is used to control how ArcCAD creates entity feature links.

Link Mode

The three options available for Link Mode are:

- 0 duplicates entities but does not create entity feature links.
- 1 duplicates entities and links the most recent entity to the FAT.
- 2 replaces entities and links to the FAT.

The default Link Mode is 2. The Link Mode setting is controlled by the Link command found on the Features menu.

Before continuing with the display capabilities of ArcCAD, let's take a look at how ArcCAD creates AutoCAD entities and how entity feature links work.

At this point the drawing file should have the street centerlines displayed on the STREETS layer. Use the Reselect command to select a subset of the street features. The specific features you are looking for are the centerlines with a class of 5.

Choose Reselect from the Query menu.

```
Command: RESELECT
Theme name (?/<POLYS>): STREETS
Reselect by?
  (Circle/ENtity/Expression/Polygon/Window): E
```

The Reselect Expression dialog box.

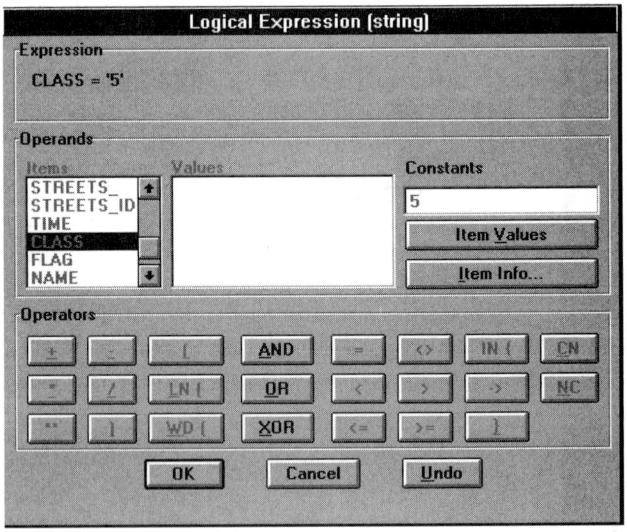

Double-click on Class in the operands, single-click on =, and type 5 in the Constants edit box. Press Return to complete the expression and select OK.

```
" CLASS = '5'" C:\INARC\STREETS arcs : 44 of 70
  selected.
```

With the class 5 centerlines selected, create a new layer named Class-5, set the color to 5, blue, and the linetype to phantom. Now execute the Arclines command to draw the features into the drawing.

```
Command: LA
LAYER ?/Make/Set/New/ON/OFF/Color/Ltype/Freeze/Thaw
  /LOck/Unlock: M
New current layer <STREETS>: CLASS-5
?/Make/Set/New/ON/OFF/Color/Ltype/Freeze/Thaw/LOck/
  Unlock: C
Color: 5
Layer name(s) for color 5 (blue) <CLASS-5>: ENTER
?/Make/Set/New/ON/OFF/Color/Ltype/Freeze/Thaw/
  LOck/Unlock: LT
Linetype (or ?) <CONTINUOUS>: PHANTOM
Layer name(s) for linetype PHANTOM <CLASS-5>: ENTER
?/Make/Set/New/ON/OFF/Color/Ltype/Freeze/Thaw/LOck/
  Unlock: ENTER
Command: ARCLINES
Theme name (?/<STREETS>): ENTER
Display using? (Properties/Item/<Symbol>): ENTER
Line symbol number (0-100) (?/<22>): 0
```

Inspection of the drawing will show that the blue centerlines have replaced the red centerlines in the drawing. This is due to the fact that Link Mode is set to the default value of 2.

Now set the Link Mode value to 1. From the Features menu, select the Link command.

```
Command: LINK
Option? (Drop/<Mode>): ENTER
New value <2>: 1
```

Use the Arclines command to redisplay the class-5 centerlines into the drawing. This time, rather than using line symbol number 0, use line symbol number 2.

```
Command: ARCLINES
Theme name (?/<STREETS>): ENTER
Display using? (Properties/Item/<Symbol>): ENTER
Line symbol number (0-100) (?/<22>): 2
```

The drawing now will show that the yellow class-5 centerlines have been written over the blue centerlines as opposed to replacing them.

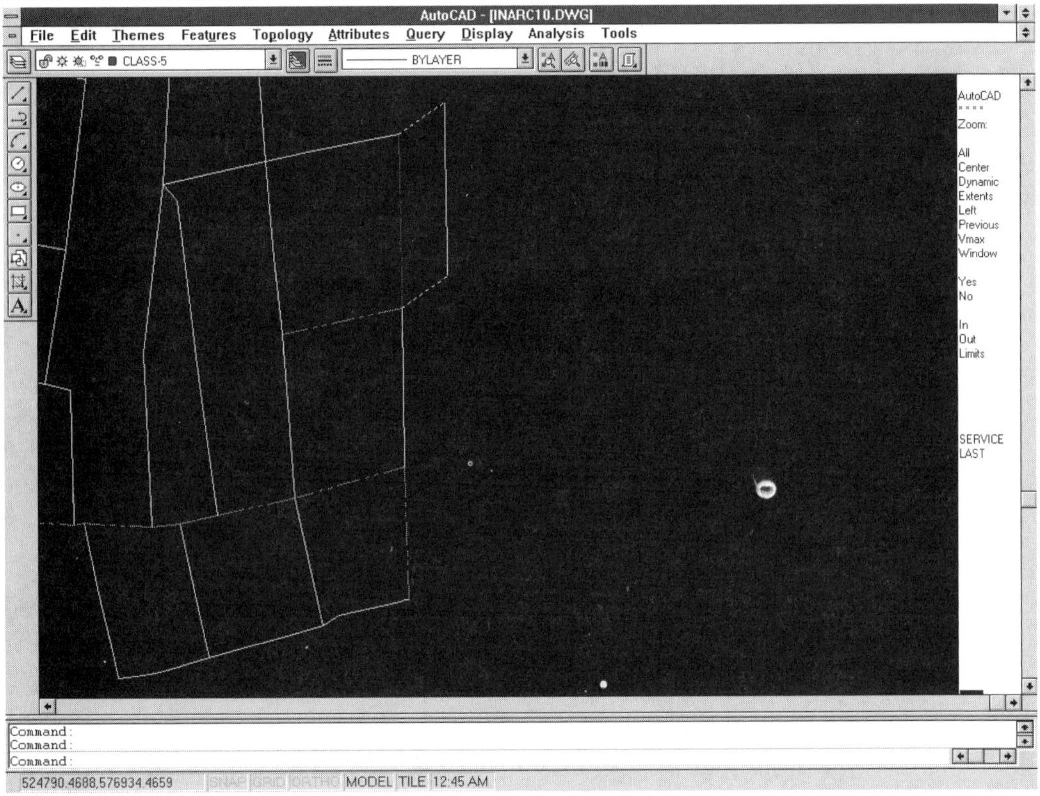

The drawing with doubled centerlines.

To demonstrate how the entity link has changed, move one of the yellow lines off of a blue line as shown above. Execute the Identify command found on the Query menu. First select the blue line.

```
Command: IDENTIFY
Theme name (?/<STREETS>): ENTER
Select an entity: <PICK THE BLUE CENTERLINE>
No feature attributes were found for the selected
  entity.
```

NOTE: *Notice that ArcCAD responds that no link information was found for the blue centerline.*

```
Select an entity: (PICK THE YELLOW CENTERLINE)
C:\INARC\STREETS arcs : 1 of 70 selected.
```

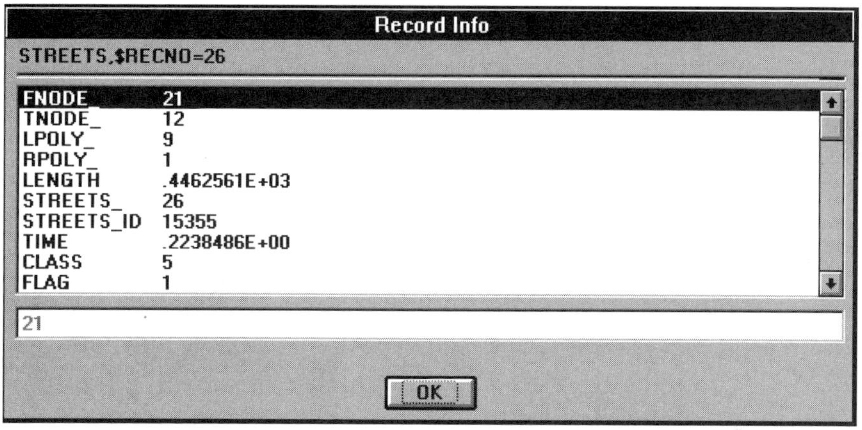

The Attributes for the selected feature.

```
Select an entity: ENTER
```

This example demonstrates that ArcCAD has removed the entity feature link to the blue centerline and created a link to the yellow centerline.

Setting the Link Mode to 0 would cause ArcCAD to duplicate entities, but no entity feature links will be created.

Because a Link Mode of 2 is appropriate for most applications, set the Link Mode back to the default of 2. Execute the Link command found on the Features menu.

```
Command: LINK
Option? (Drop/<Mode>): ENTER
New value <1>: 2
```

You should now have a good understanding of how ArcCAD creates AutoCAD entities from features and the method used to create links to these AutoCAD entities. Let's continue and take a look at the display commands for each theme type.

Displaying Lines

The Lines menu under Display contains commands for displaying line theme features. These commands allow the display of line features based on a symbol number entered at the symbol prompt, with a symbol based on an item in the database or with a symbol based on a lookup table. You can also display line features labeled with the User-ID. Additional commands are available for labeling line features with an item from the attribute table or a lookup table. The Display Line menu is shown below.

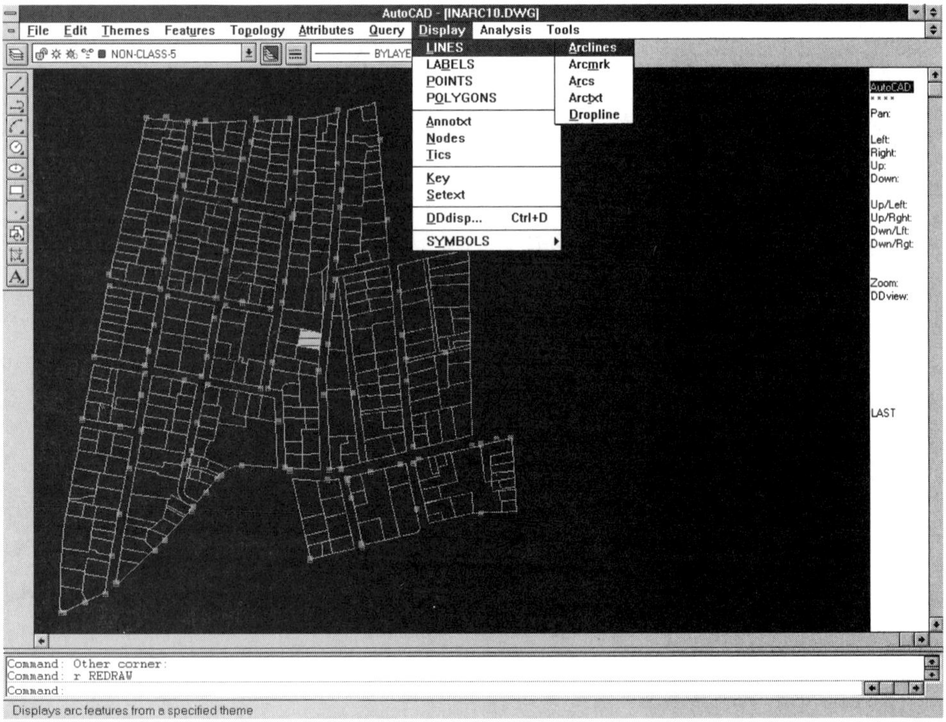

The Display Lines menu.

In the previous section you were introduced to the Arclines command and its ability to display features based on a symbol number. Arclines can also be used to display features based on an item in the database or to supply the AutoCAD properties of the features based on input or an item in the database.

Displaying theme features based on an item in the database allows ArcCAD to control feature symbology by using a numeric item in the database. The sample data for the street centerlines contains a numeric item named CLASS. The value of the CLASS item is 2, 4, or 5. This item can easily be used to thematically map the centerlines.

To thematically map the centerlines, execute the Clearsel and then the Arclines commands.

```
Command: CLEARSEL
Command: ARCLINES
Theme name (?/<STREETS>): ENTER
Display using? (Properties/Item/<Symbol>): I
Item (?): CLASS
Optional Lookup table (?/<none>): ENTER
```

Using Arclines in this manner tells ArcCAD to walk through each record, displaying lines with the symbol number contained in the class item.

Arclines can also display lines using AutoCAD properties. The properties that can be controlled by the Arclines command are Layer, Color, Linetype, Linewidth, Elevation, Thickness, and Curve. Each of the property values may be entered at the command line or extracted from an item value contained in the FAT.

In the last example you used the CLASS value to control the color of the lines being displayed. Let's try a similar example, this time both the color and the Line width will be determined by the CLASS item in the FAT.

```
Command: ARCLINES
Theme name (?/<STREETS>): ENTER
Display using? (Properties/Item/<Symbol>): P
Layer/Item/<CLASS-5>: L
Layer: STREETS
Color/Item/<BYLAYER>: I
Item (?): CLASS
Linetype/Item/<BYLAYER>: ENTER
```

```
Linewidth/Item/<0.0>: I
Item (?): CLASS
Elevation/Item/<0.0000>: ENTER
Thickness/Item/<0.0000>: ENTER
Curve item name (?/<none>): ENTER
```

The input above tells ArcCAD to display the class 5 lines in symbol number 5 with a width of 5, class 4 with a symbol number of 4 and a width of 4, and class 2 with a symbol number of 2 and a width of 2. This is one of ArcCAD's most powerful display features.

A more complex example of the properties' capabilities can be seen in displaying a pipeline coverage. The pipes for the lines could be displayed as 3D objects at the correct elevation and size, all based on attributes in the FAT. After the display, a complete schematic of the pipe system can be seen by rotating the AutoCAD drawing into perspective.

With the centerlines displayed in the drawing based on attributes in the table, the Arctxt command may now be used to label the centerlines with their street names.

Arctxt has three options (Two, Line, <One>). These options control whether the text is displayed in single line, stacked, or aligned with the feature being labeled.

To label the street centerlines, set the text size for the labels with the Textsize command and execute the Arctxt command.

```
Command: TEXTSIZE
New value for TEXTSIZE <0.2000>: 15
Command: ARCTXT
Theme name (?/<STREETS>): ENTER
Item (?): NAME
Optional lookup table (?/<none>): ENTER
Method? (Two/Line/<One>): L
Text always right side up? (No/<Yes>): N
Alignment? (Center/Middle/Right/TL/TC/TR/ML/MC/
  MR/BL/BC/BR/<Left>): M
```

The drawing should look like the following figure.

The drawing with centerlines labeled using the Line option.

Displaying Polygons

The Polygons menu found under the Display pulldown contains all of the commands for displaying, labeling, and shading polygons.

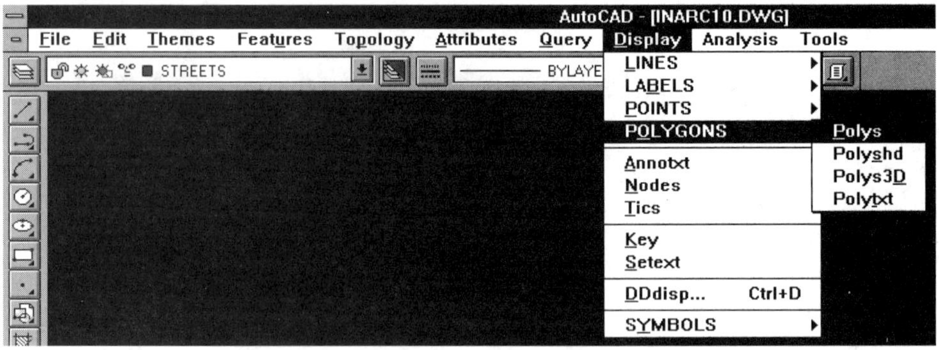

The Display Polygons menu.

ArcCAD handles the display of polygons similarly to that of Lines. When displaying polygons, you may draw either their boundaries or shade them with hatch patterns. The hatch patterns available may be displayed by entering a question mark at any Shade symbol number prompt.

The Polys command displays polygon boundaries. This command has been used several times in previous chapters. The only input to the command is the Line Symbol Number to use for the display. When displaying polygons using Polys, entity feature links are not created for the polygon boundaries. In order to edit polygons, line themes must be defined and displayed for the polygon coverage, thereby creating entity feature links for the boundaries.

The Polyshd command is used to shade polygons with hatch patterns from the Shade Symbol Set. The Polyshd command supplies the same list of options as the Arclines command.

To see how the Polyshd command works, execute it now.

```
Command: POLYSHD
Theme name (?/<PARCELS>): ENTER
Display using? (Properties/Item/<Symbol>): ENTER
Shade symbol number (0-100) (?/<0>): ?
```

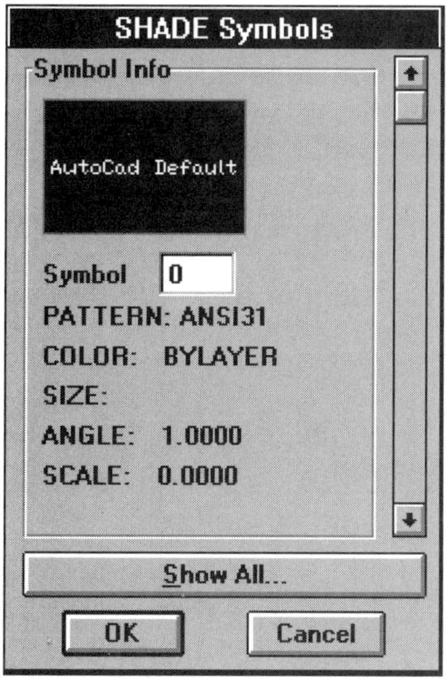

The SymIndex dialog box for shade symbols.

The Shade Symbol dialog box is similar to the Line Symbol dialog box. Entering a number in the symbol number edit box will display that symbol in the image tile. To scroll through the various symbols, select the down arrow along the right side of the dialog box. The image tile updates as the slider scrolls down. Press the Show All button at the bottom of the dialog box to show 25 of the shade symbols at a time.

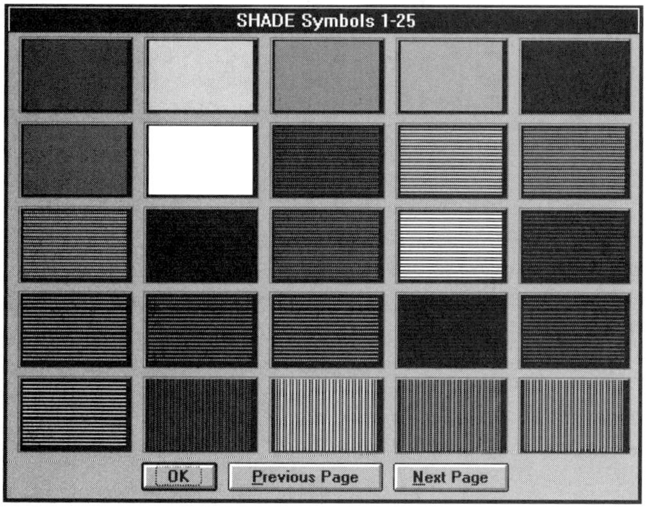

The SymInfo subdialog box for shade symbols.

The Show All button allows you to move a page at a time through all of the shade symbols. Once the desired symbol is found, simply double-click on it to return to the Shade Symbol dialog box. At this time, select OK from the bottom of the dialog box. In the Shade Symbol dialog box, type 55 in the edit box. A magenta diagonal hatch should display in the image tile. Select OK from the bottom of the dialog box.

The drawing should look like the following figure.

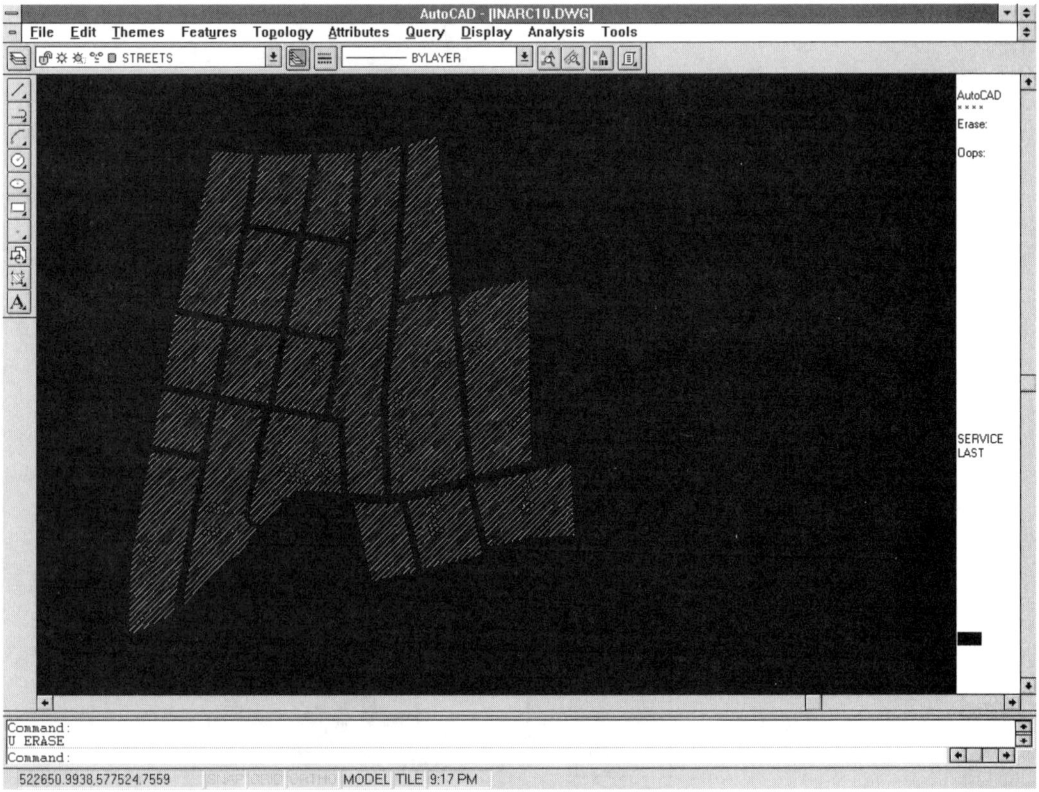

The drawing file with parcels shaded.

As you can see from looking at the drawing, the parcels have been shaded but their boundaries have not been drawn. To add the polygon boundaries to the drawing, use the Polys command.

The drawing file with parcels shaded and outlined.

The Properties option of the Polyshd command works similarly to that of the Arclines command. The difference is that the Polyshd properties provide prompts appropriate for hatch patterns as opposed to line properties.

```
Command: POLYSHD
Theme name (?/<PARCELS>): ENTER
Display using? (Properties/Item/<Symbol>): P
Layer/Item/<STREETS>: ENTER
Color/Item/<BYLAYER>:  ENTER
Angle/Item/<0.0>: ENTER
Pattern/Item/<line>: ENTER
Elevation/Item/<0.0000>: ENTER
Thickness/Item/<0.0000>: ENTER
Scale/Item/<1.0>: ENTER
```

Text Symbols

The Polytxt command is used to display attribute values in the drawing. The only options to the Polytxt command are the Item and the text symbol number.

This is an appropriate time to discuss text symbols and the Textsize command. ArcCAD handles the display of text in two different ways. If text symbol number 0 is used, the text is displayed on the current layer and color, in text style Standard, with a Font File of TXT. The height of the text is controlled by the Textsize setting. If any symbol number other than 0 is used, the text is displayed on the current layer, with the color and font defined in the symbol set. The size of the text is determined by multiplying the size set in the symbol set by the current zoom magnification.

To demonstrate how ArcCAD handles text display using symbol numbers other than 0, execute the Polytxt command.

```
Command: POLYTXT
Theme name (?/<PARCELS>): ENTER
Item (?): TOTVAL
Optional lookup table (?/<none>): ENTER
Text symbol number (0-100) (?/<0>): ?
```

The SymInfo dialog box Text Symbols.

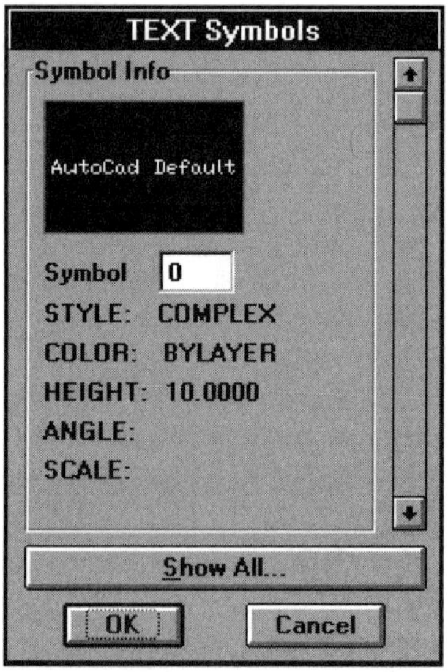

The Text Symbol dialog box works identical to the Line and Shade Symbol dialog boxes.

Select the Show All button at the bottom of the dialog box.

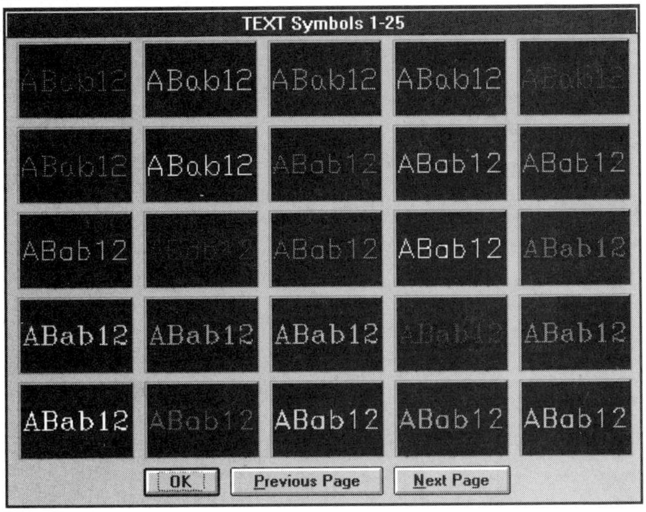

The SymIndex dialog box for Text Symbols.

From the page display double-click on symbol number 18 and select OK from the Text Symbol dialog box.

The text is displayed into the drawing with its size calculated based on the zoom magnification multiplied by 0.0150 (the number displayed as the Height in the Text Symbol dialog box). With the drawing zoomed out so that the entire work area is shown, the text is much too large.

The drawing with polytxt displayed.

To change the size of this text, execute the AutoCAD Undo command to remove the text. Zoom in to the view shown below.

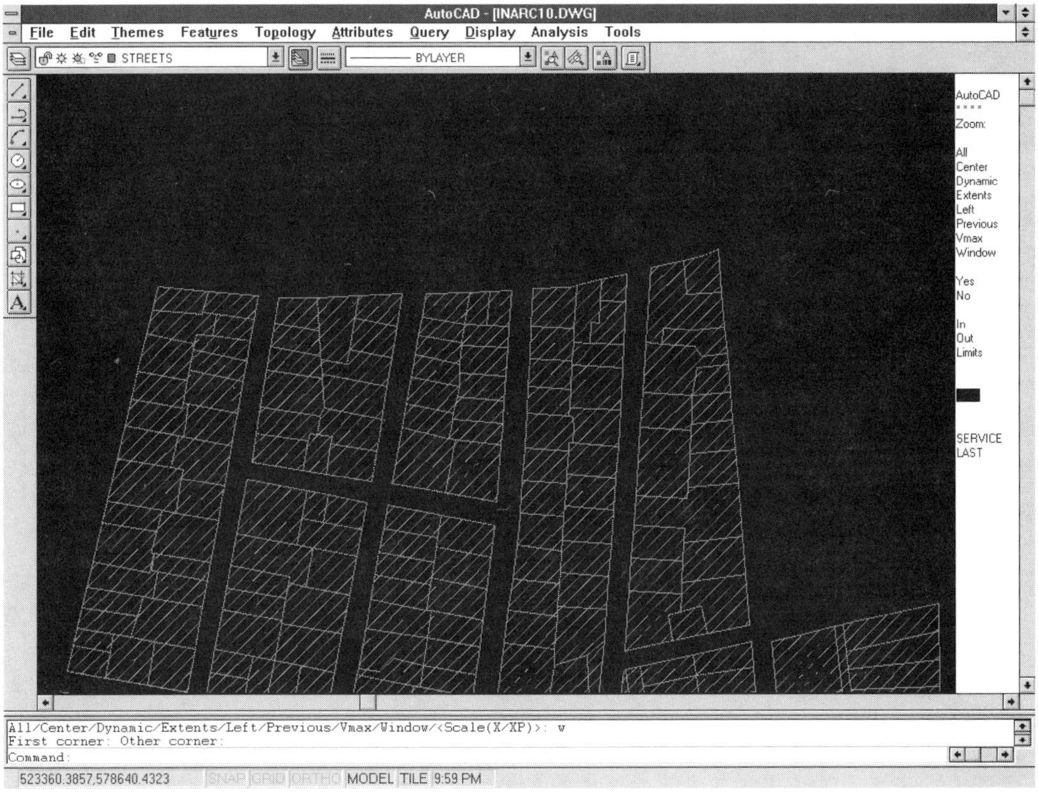

The drawing zoomed in.

Reexecute the Polytxt command. In the symbol number edit box type `18`, press Enter and select OK.

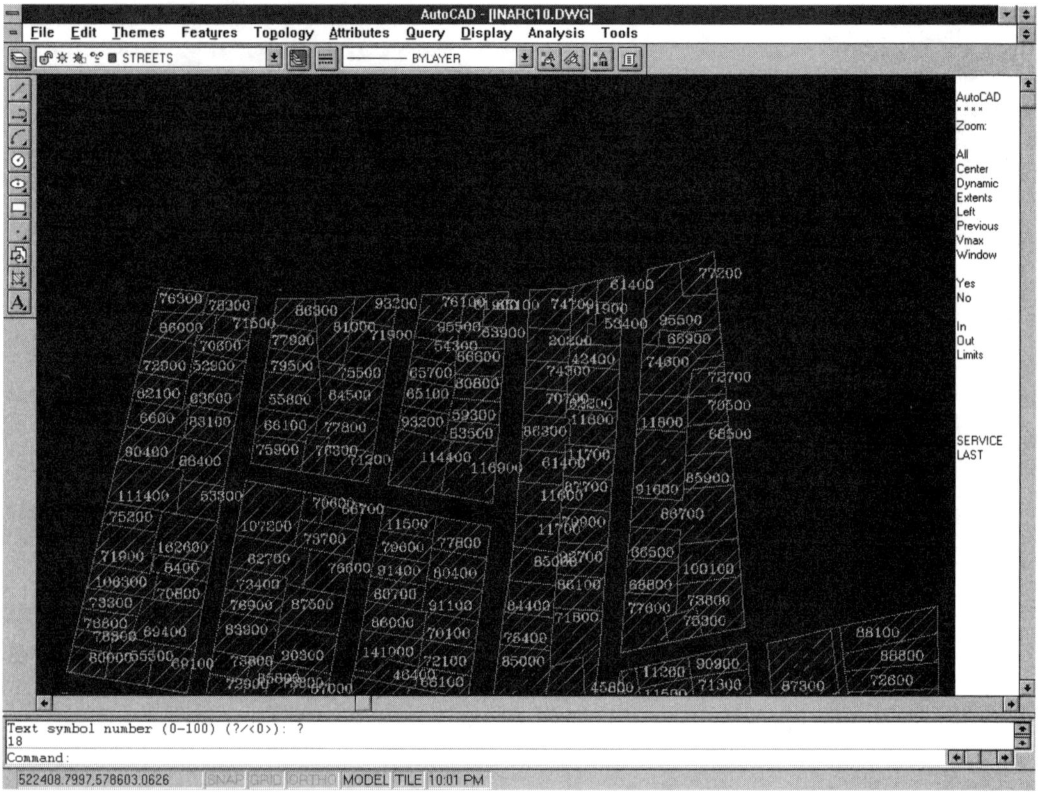

The drawing with resized polytxt.

The text in the drawing is now drawn with a smaller height. This feature is useful when creating multiple maps from a series of themes. Each map can be at a different scale. To change scales, simply zoom in or out and redisplay the information.

The last command to be covered in this section is the Polys3D command. Polys3D is a powerful, thematic mapping tool, allowing attribute values to be presented as the height of 3D objects in AutoCAD. The options to the Polys3D command are the same as that of the Arclines command.

An example of the Polys3D command is shown below. In this example, the value of the SQFTVAL item is displayed as the thickness of the parcels. Once the polys are displayed, the drawing can be rotated to an isometric or perspective view and shaded.

Before trying Polys3D, use the AutoCAD Erase command to clear the drawing. Next, execute the Polys3D command.

```
Command: POLYS3D
Theme name (?/<PARCELS>): ENTER
Display using? (Properties/<Symbol>): ENTER
THickness/Item/<0.0000>: I
Item (?): SQFTVAL
Line symbol number (0-100) (?/<0>): 2
```

With the polys displayed in the drawing, use the AutoCAD Vpoint command to rotate to an isometric view, set the Shadedge variable to 1, and shade the drawing.

```
Command: VPOINT
Rotate/<View point> <0.0000,0.0000,1.0000>: -1,-1,.25
Regenerating drawing.

Command: SHADEDGE
New value for SHADEDGE <3>: 1

Command: SHADE
Regenerating drawing.
Shading complete.
```

The drawing should look like the following figure.

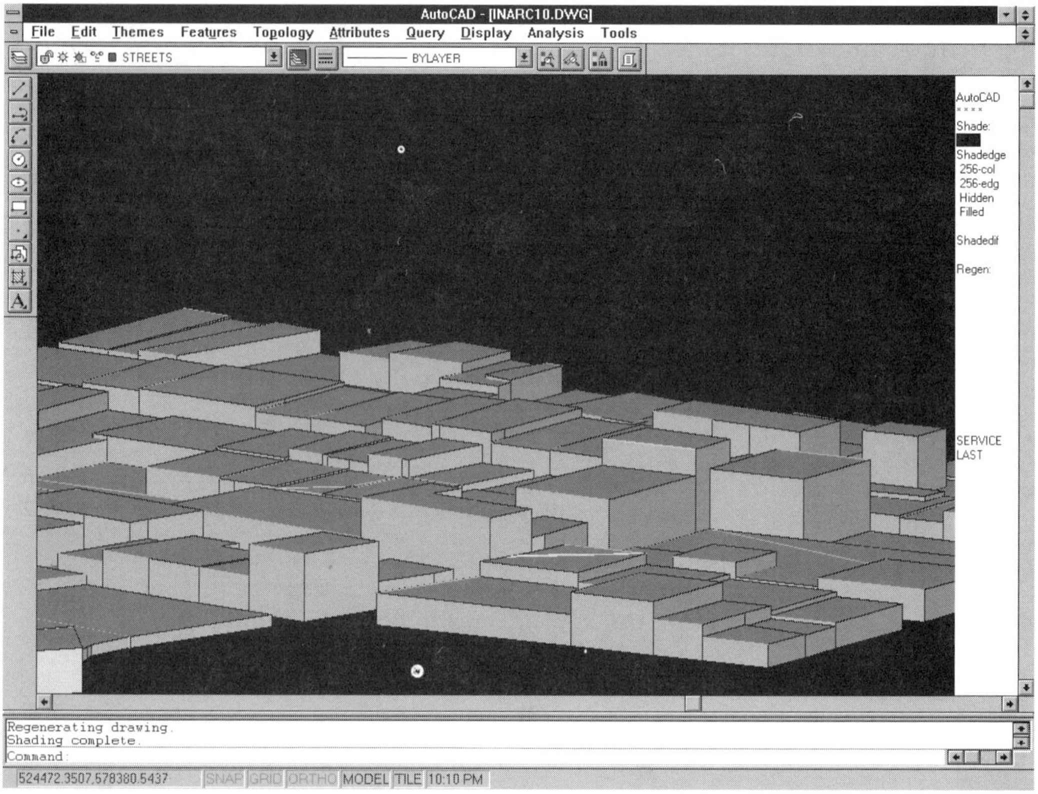

The Parcels displayed with Polys3D.

The properties option of the Polys3D command allows the control of the following AutoCAD properties: Layer, Color, Linetype, Linewidth, Elevation, and Thickness. Each of these properties can be specified at the command prompt or by using the value of an item in the FAT.

> Before moving on, execute the AutoCAD Plan command and press enter at the default of <current UCS>. Once back in plan view, erase all the drawing entities.

There are two additional commands that, although not found on the Polygons menu, should be discussed here. They are the Labelmrk and Labeltxt commands. These two commands can be found on the Display pulldown menu under the Label pullout.

The Labelmrk command allows polygon label points to be displayed as point markers.

The Labeltxt command is similar to the Polytxt command described previously. The one notable difference between the two commands is that Labeltxt allows the specification of the text justification using any of the AutoCAD text justification options.

Displaying Points

The Display Points menu has three commands. Each of these commands work similar to their respective commands on the Lines and Polygons menus.

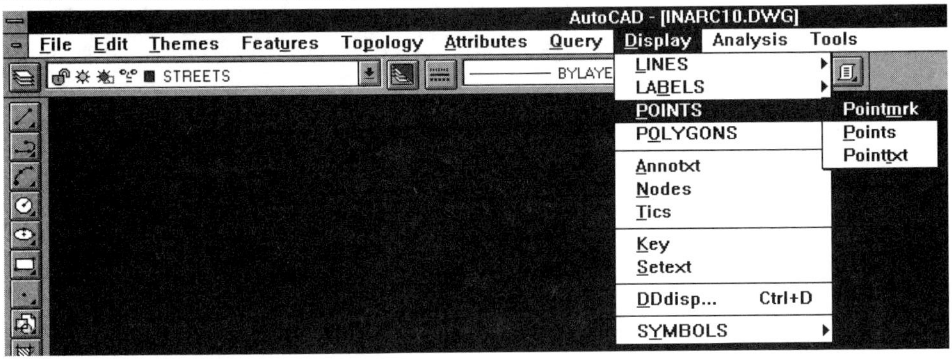

The Display Points menu.

Although the commands on the Points menu work similar to commands discussed earlier, there are some important topics to cover with regard to points display.

When displaying point features into drawing files, ArcCAD uses a series of blocks that are defined when ArcCAD loads. These blocks have the name of ESRI_M1 through ESRI_M16. To view these blocks, select MrkIndex from the SYMBOLS menu found on the Display pulldown menu.

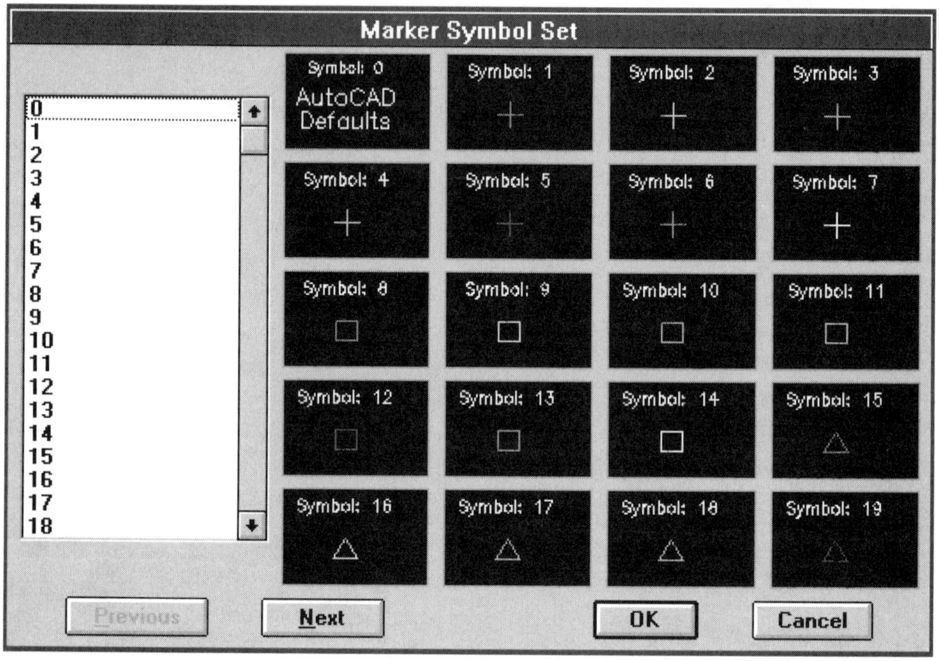

The SymInfo dialog box for Marker Symbols.

The definition of each of these symbols may be displayed by using the SymInfo command also found on the Symbols menu.

```
Command: syminfo
Symbol type? (Line/Marker/Shade/Text): M
```

The SymIndex dialog box for Marker Symbols.

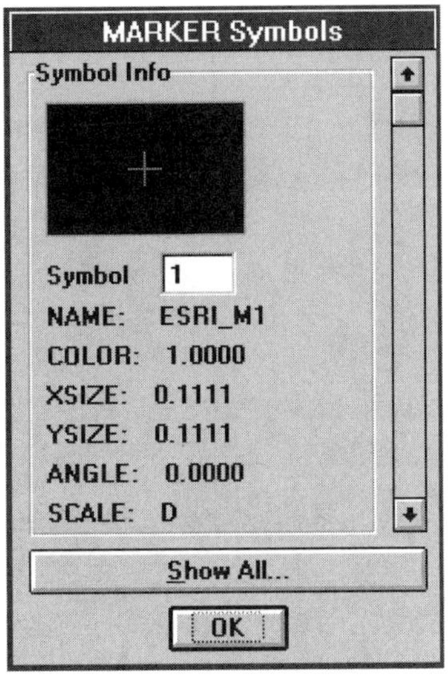

The Pointmrk command from the Points menu allows these symbols to be used when displaying point features in the drawing. The symbol to be used may be specified by entering a symbol number at the command prompt, by using an item value from within the FAT, or by looking up an item value in a look-up table. An example of the Pointmrk command using an item value is shown below.

```
Command: POINTMRK
Theme name (?/<PARCELS>): UTILS
Display using? (Properties/Item/<Symbol>): I
Item (?): UTILS_ID
Optional lookup table (?/<none>): ENTER
```

The UTILS_ID item was used to supply a unique symbol number for each feature in the UTILS theme. The drawing should now contain a series of unique symbols, each one representing a hydrant or light location. The drawing file should look like the following figure.

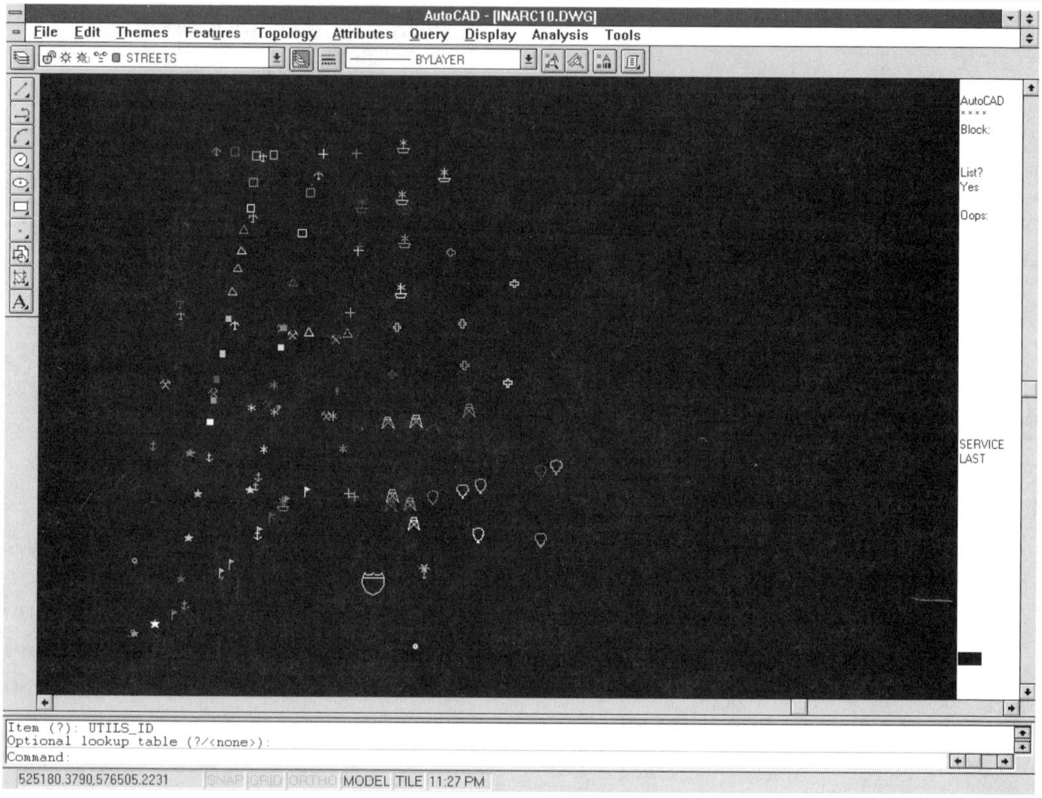

The drawing file with marker symbols displayed.

ArcCAD handles the display size of symbols the same as text. As the zoom magnification of the drawing increases, so will the symbol sizes. If the zoom magnification decreases and the points are redisplayed, their size will decrease.

The Pointtxt command will allow the display of item values displayed as text. Pointtxt offers the same options as the Polytxt command shown earlier.

The Points command will allow point locations to be plotted on the drawings with the default symbol. The command will prompt to automatically label the points with the User-ID. Displaying the points with IDs simplifies the process of populating a database with item values as described in Chapter 4.

In the next chapter you will use a look-up table to display the hydrants and lights in the creation of a complete thematic map.

Review

You have been introduced to ArcCAD's method of displaying lines, polygons, points, and text in drawing files. Using the commands described previously, you now have the ability to create drawing files by using query commands to select specific features and display them using various symbology.

In the next chapter you will learn to automate the process of creating thematic maps through the use of look-up tables and item values to specify symbol numbers. You will also create keys (legends) for your maps and add north arrows and title blocks.

Chapter 12

Thematic Mapping

Creating Intelligent Maps in AutoCAD

Thematic mapping is the process of using color, line type, line weight, and patterns to describe features on a map. ArcCAD allows you to set these properties based on items within the feature attribute tables (FATs) and to look up feature attribute values in record themes. Once the information is displayed in AutoCAD, the Save As or WBlock command can be used to save the current display as a drawing. The screen can then be erased and a new set of features with a new set of properties can be displayed and saved as a second map. This process allows AutoCAD and ArcCAD to be used in a thematic mapping environment.

In this chapter you will create a series of maps and save them as individual AutoCAD drawing files.

Setup

To begin this chapter, load AutoCAD and open the INARC12 drawing file.

This drawing file has the PARCELS, LANDUSE, STREETS, ULINE, TREES, and TREELUT themes defined. A layer for each theme has been created and the extents of the drawing have been set.

The drawings to be created include a map of the street centerlines with symbology based on the class and labeled with the street name. A map displaying the location and crown width of trees overhanging utility lines will also be created. The last map will be of the parcels. On this map, the parcels will be shaded with a symbol from a lookup table. The lookup item will be the year the parcel was developed. You will create a legend for each of the maps.

Creating a Street Centerline Map

The first map you will create uses the STREETS theme. Although it is the simplest of the three, the command sequence is similar for the other two maps.

In Chapter 11 you saw a brief example of how the street centerlines might be thematically mapped based on an attribute in the FAT. In that example, all of the centerlines were displayed in red. The Reselect command was used to select all centerlines with a class of 5, which were displayed in blue. A key could have been created specifying that the blue centerlines were class 5 and the red were all other classes. Although this was indeed a thematic map, there is a much easier way to create it.

In this example, you will display the centerlines using symbology that is derived directly from the CLASS value in the FAT. This eliminates the need to build selection sets.

> To begin, be sure the STREETS layer is current. Select the Arclines command found on the Display menu and follow the input below.

```
Command: ARCLINES
Theme name (?/<theme>): STREETS
Display using? (Properties/Item/<Symbol>): I
Item (?): CLASS
Optional Lookup table (?/<none>): ENTER
```

By using the Item option in the Arclines command, you have told ArcCAD that rather than specifying a symbol number at the command prompt, you want the symbol number for each feature to come directly from the FAT. In the drawing the class 5 streets are displayed in symbol number 5, the class 4 streets in symbol number 4, and the class 2 in symbol number 2. The drawing file should look as shown below.

The drawing file with centerlines displayed based on CLASS.

The ability to display features based on items in the database, although powerful, can at times be limiting. For example, this can only be done when the item defining the symbols is numeric. The value also has to define acceptable symbols for the features being displayed. In the case of the street centerlines, the class values define yellow, blue, and cyan lines. You may decide that blue and cyan are not acceptable for your map. To solve this problem you can create a lookup table to define the symbols to use. With a lookup table, the street class will be looked up in the table and the corresponding symbol number will be used to display the centerline.

You will have the opportunity to use lookup tables later in this chapter.

The next step in creating the centerlines map is to add the street names to the drawing. Before bringing the street names into the drawing, change the current layer to STREET_TEXT. Set the text size and execute the Arctxt command.

```
Command: TEXTSIZE
New value for TEXTSIZE <0.0000>: 10
Command: ARCTXT
Theme name (?/<STREETS>): ENTER
Item (?): NAME
Optional lookup table (?/<none>): ENTER
Method? (Two/Line/<One>): LINE
Text always right side up? (No/<Yes>): N
Alignment?
  (Center/Middle/Right/TL/TC/TR/ML/MC/MR/BL/BC/BR/
  <Left>): M
Text symbol number (0-100) (?/<0>): ENTER
```

Creating Legends

One of the most important pieces of a thematic map is the legend or key. The key is a series of symbols with text identifying what each of the symbols represents. Using the key, anyone reading the map can immediately determine the contents and meaning of the map. You can create an ArcCAD key file using any text editor. These files can then be read into AutoCAD as complete legends.

Once the text has been added to the drawing, you can create a legend for the map, using a text editor.

If you are using AutoCAD for Windows, use the Notepad to create the key file. If you are using AutoCAD for DOS, use the Shell command to enter DOS and use the Edit command to create the legend. The legend file should look like the following figure.

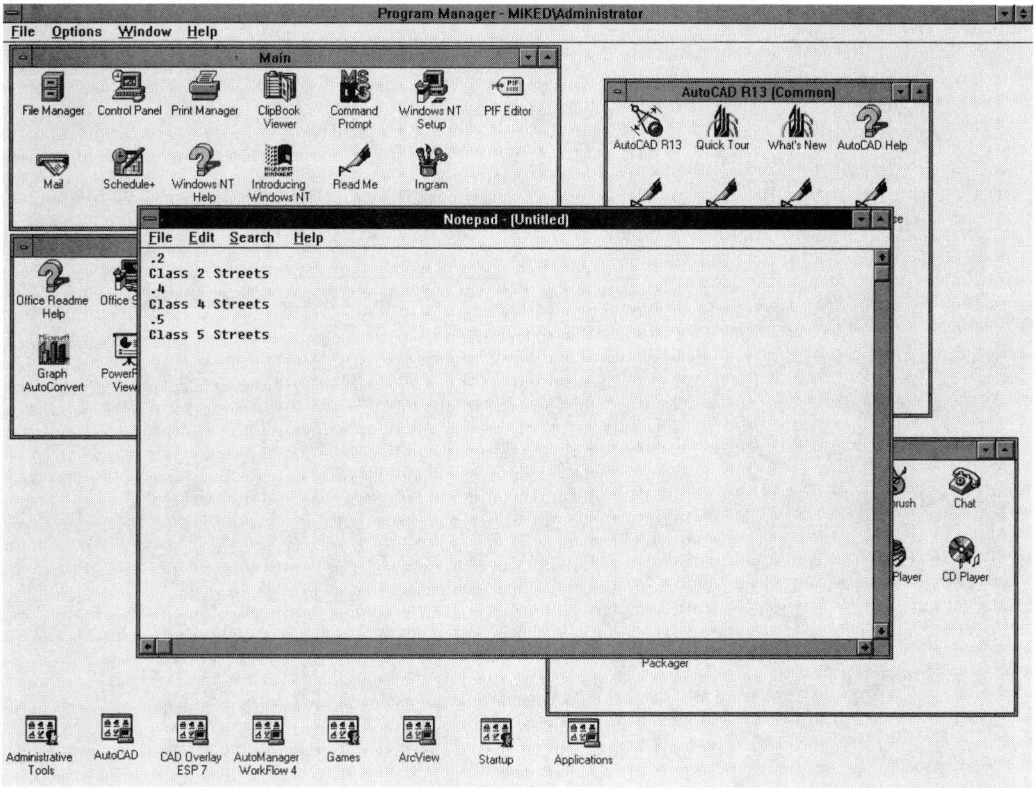

The STREETS map key file.

Once the input shown above has been entered, save the file as STREETS.KEY. The filename is user defined, but the extension must be .KEY.

Once the file has been saved, return to the drawing file. To create a legend from the key file, use the Key command found on the Display menu and follow the input below.

```
Command: TEXTSIZE
New value for textsize: <10.0>: 40

Command: KEY
Symbol class? (Line/Marker/Shade): L
Key file name: STREETS.KEY
Width of key box <0.7500>: 100
```

```
Height of key box <0.7500>: 75
Horizontal separation <0.2500>: 25
Vertical separation <0.2500>: 25
Draw key box outline? (No/<Yes>): YES
Key upper left corner (X,Y): SELECT A POINT
```

The Key command allows the creation of Line, Marker, and Shade legends. The STREETS legend should look like the following figure.

The STREETS legend.

The last step in the development of the centerlines map is to use the AutoCAD Insert command to add a title block, north arrow, scale, and any notes to your map. As of now the map should look like the following figure.

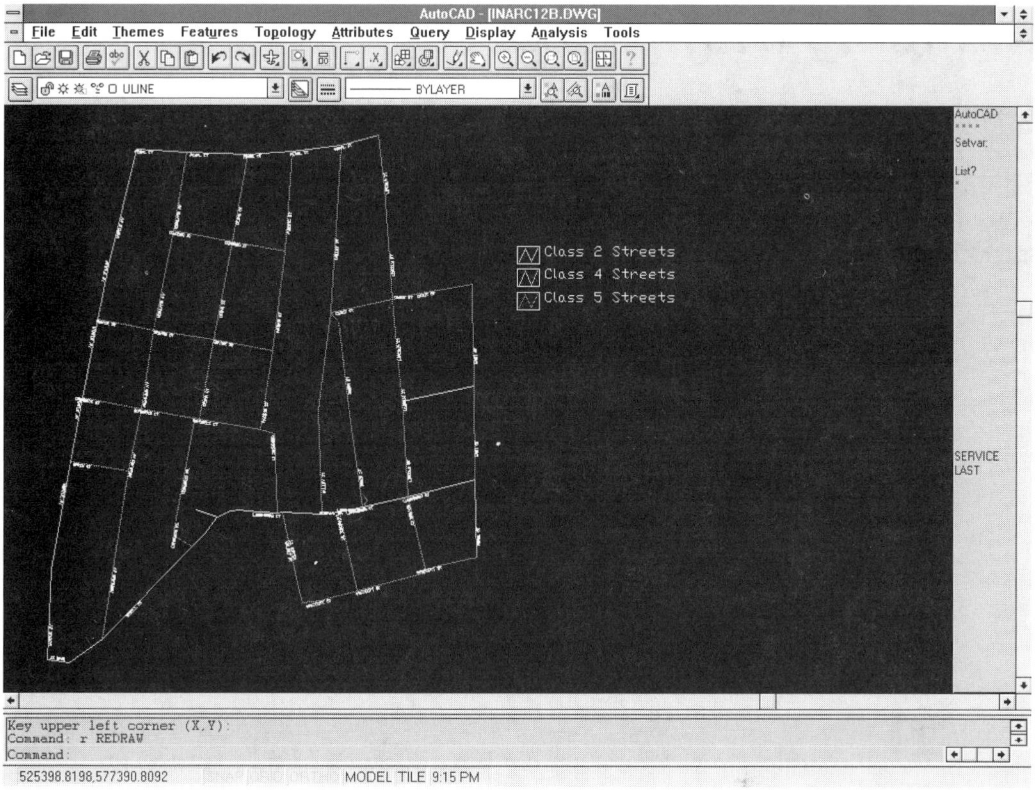

The STREETS map.

> Once all of the map components have been added, the map can be saved to a new AutoCAD drawing file by using the AutoCAD Wblock or the Saveas commands. If you would like to save this drawing, execute the Wblock command.

This completes the creation of the STREETS map. The next map you will create will be a little more complex, using lookup tables to specify display symbols. To begin the UTILITY map, use the AutoCAD Erase command to clear the screen if necessary.

Mapping Utility Lines

The process of creating the UTILITY map will include several calculations to determine which trees possibly interfere with the power lines. You have seen one method of determining which trees interfere in chapter 9. Here you will learn a new set of commands that allow you to created similar information.

To begin, set the ULINE layer current and display the power lines into the drawing.

```
Command: LA
LAYER
  ?/Make/Set/New/ON/OFF/Color/Ltype/Freeze/Thaw/LOck/
  Unlock: S
New current layer <STREETS>: ULINE
?/Make/Set/New/ON/OFF/Color/Ltype/Freeze/Thaw/LOck/
  Unlock:
Command: ARCLINES
Theme name (?/<THEME>): ULINE
Display using? (Properties/Item/<Symbol>):
Line symbol number (0-100) (?/<0>): 2
```

With the utility lines now displayed in yellow, add the trees to the drawing.

```
Command: POINTMRK
Theme name (?/<ULINE>): TREES
Display using? (Properties/Item/<Symbol>): ENTER
Marker symbol number (0-100) (?/<0>): 24
```

The next task is to determine which of the trees overhang the utility lines, which might cause potential problems. To do this you will determine the distance between the trees and the utility lines using the Near command. Next, you will divide the crown width item in the TREES FAT by 2 using the Calculate command. You will then calculate a new item in the database that will show which of the trees overhang the utility lines. This item will be calculated by subtracting the half width of the trees from the distance to the utility lines. A positive number will indicate a tree overhangs the utility line.

The first step in this process is to use the Near command to calculate the distance between the trees and the utility lines.

```
Command: NEAR
Input theme (?/<trees>): ENTER
Near theme (?): ULINE
Feature type? (Node/<Line>): ENTER
Search radius <compute>: 200
Please wait...
```

A new item named Distance has been added to the TREES FAT. This item represents the distance between the trees and the utility lines. Create a new item in the TREES FAT to contain the crowns half width.

```
Command: ADDITEM
Theme name (?/<trees>): ENTER
Item name (?): HALFWIDTH
Item width: 10
Item type?  (Character/Date/<Numeric>): ENTER
Item decimal places <0>: 2
Start item <last>: ENTER
```

Now use the Calculate command to divide the crown width by 2.

```
Command: CALCULATE
Theme name (?/<trees>): ENTER
Target Item (?): HALFWIDTH
```

The Calculate dialog box.

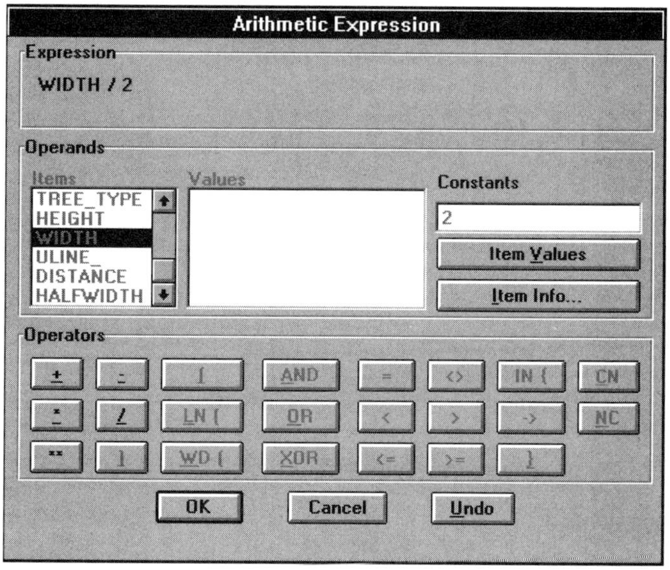

```
" WIDTH / 2"
```

The last step in the analysis portion of the utilities map is to do the calculation to determine which of the trees are within the crowns half width of the power lines.

```
Command: ADDITEM
Theme name (?/<trees>): ENTER
Item name (?): INPOWER
Item width: 10
Item type?  (Character/Date/<Numeric>): ENTER
Item decimal places <0>: 2
Start item <last>: ENTER
Command: CALCULATE
Theme name (?/<trees>): ENTER
Target Item (?): INPOWER
```

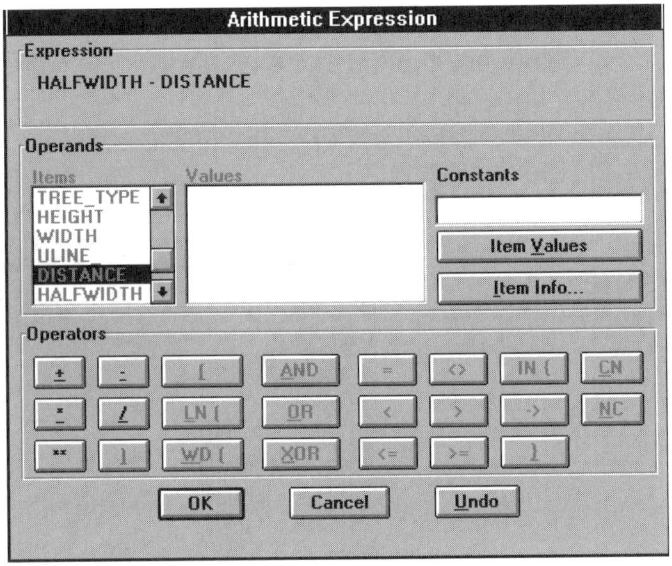

The Calculate dialog box as the distance between trees and powerlines is determined.

```
" HALFWIDTH - DISTANCE"
```

The new INPOWER item will allow a query of only features with a value greater than 0 to be selected. These features will be the trees overhanging the power lines.

```
Command: RESELECT
Theme name (?/<trees>): ENTER
Reselect by?
  (Circle/ENtity/Expression/Polygon/Window): E
```

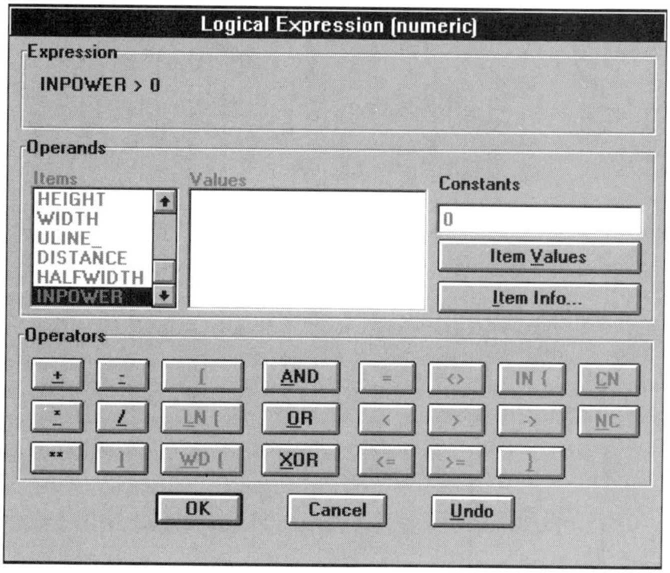

The Reselect dialog box.

```
" INPOWER > 0" C:\INARC\TREES points : 28 of 114
  selected.

Command: POINTMRK
Theme name (?/<trees>): ENTER
Display using? (Properties/Item/<Symbol>): ENTER
Marker symbol number (0-100) (?/<0>): 26
```

The drawing should now look like the following figure.

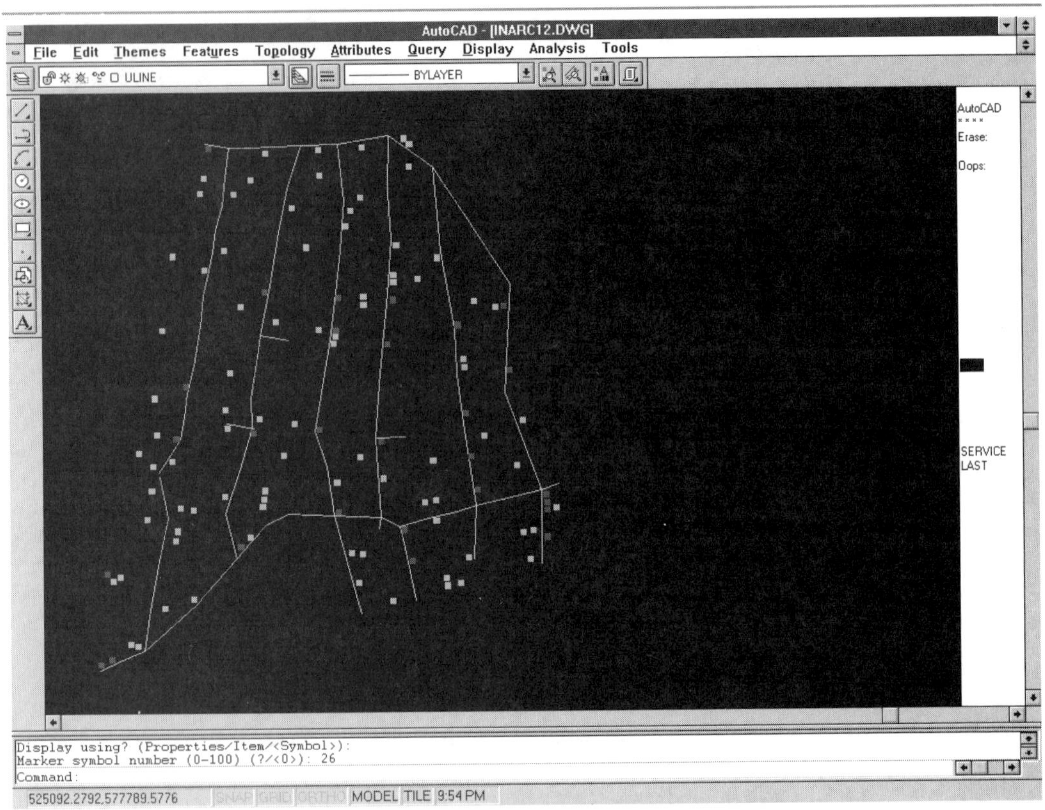

The drawing with calculations made.

You will now create a set of buffers for the trees, representing their crown width. To do so, first Extract the selection set to a new coverage. Remember, this is necessary because ArcCAD only buffers entire coverages.

```
Command: EXTRACT
Input theme (?/<TREES>): ENTER
Output theme (?): TREES2
Theme TREES2 is not defined.
```

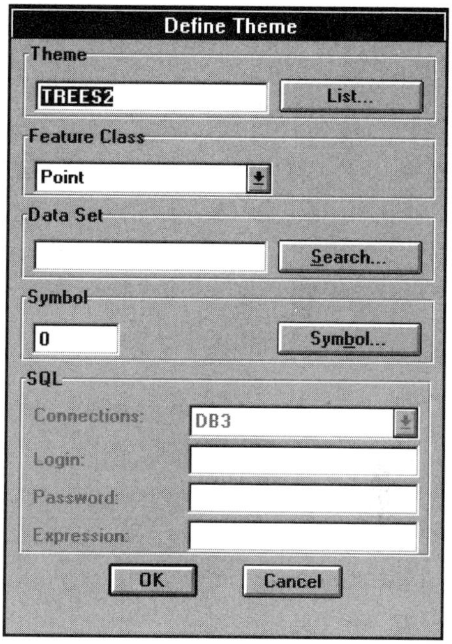

The Define theme dialog box.

```
Defining TREES2...
Extracting features from TREES to create TREES2.
28 of 114 records selected.
Extracting points...
Number of Points in Coverage...      114
Number of Points Reselected....       28
Creating Point attribute table.
```

To create the buffers for the extracted trees, execute the Buffer command. In this example you will use an item in the database to set the buffer distance.

```
Command: BUFFER
Input theme (?/<TREES2>): ENTER
Output theme (?): TREEWDTH
Theme TREEWDTH is not defined.
```

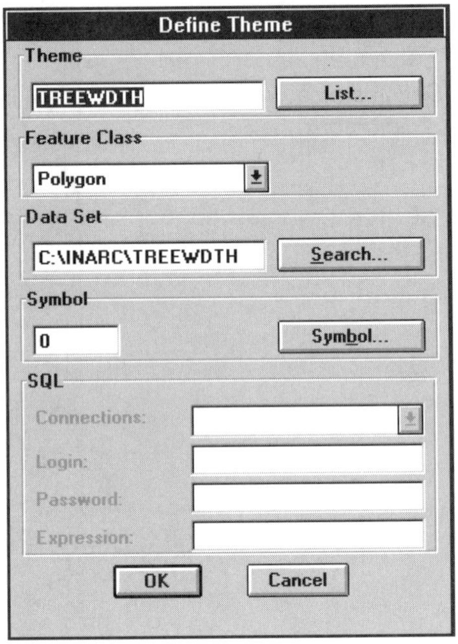

The Define Theme dialog box for tree width.

```
Defining TREEWDTH...
Buffer using? (Item/<Constant>): i
Buffer item (?): WIDTH
Optional lookup table (?/<none>): ENTER
Fuzzy tolerance <compute>: ENTER
Buffering...
```

To complete the map, use the Polyshd command to display the tree buffers, the Arclines command to display the road centerlines, and the Pointmrk command to display the light poles.

```
Command: POLYSHD
Theme name (?/<TREES2>): TREEWDTH
Display using? (Properties/Item/<Symbol>): ENTER
Shade symbol number (0-100) (?/<0>): 6

Command: POINTMRK
Theme name (?/<treewdth>): STRTLTS
Display using? (Properties/Item/<Symbol>): ENTER
Marker symbol number (0-100) (?/<0>): 48

Command: ARCLINES
Theme name (?/<STRTLTS>): STREETS
```

```
Display using? (Properties/Item/<Symbol>): ENTER
Line symbol number (0-100) (?/<0>): 43
```

The drawing file should now look like the following figure.

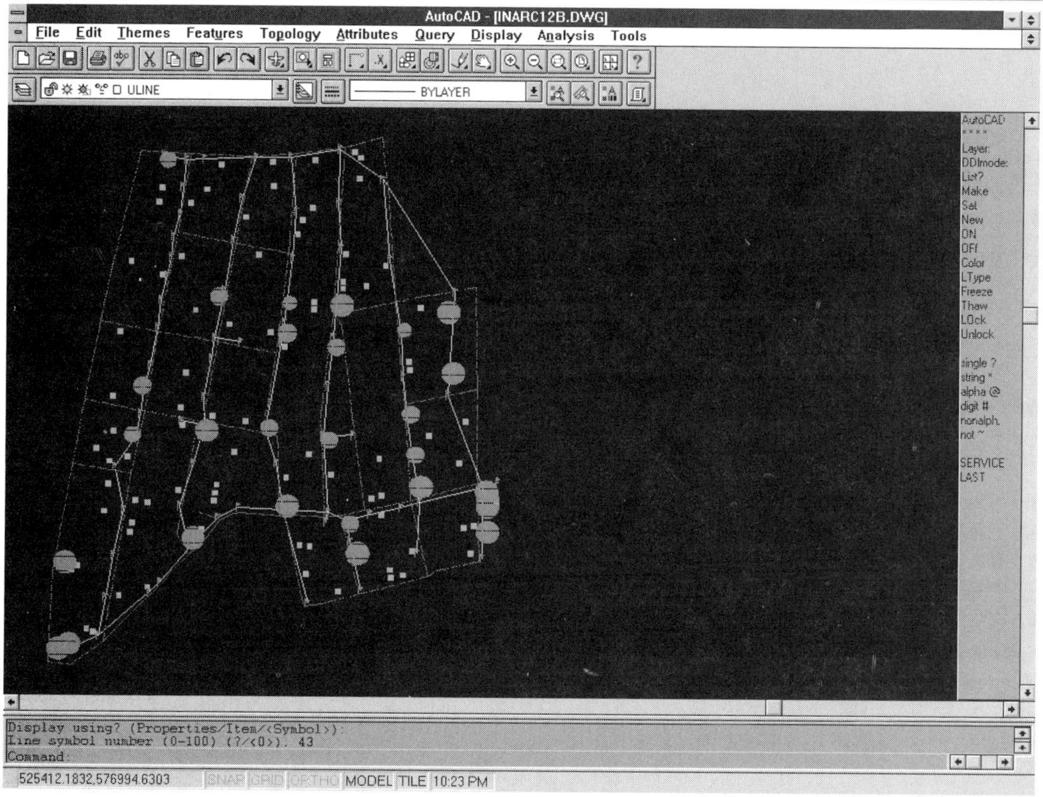

The INARC12 drawing file.

The last step in creating your map is to add a legend. For this map, three separate legends will be created. The first legend will be for the buffer shades, the second for the point symbols representing the trees and lights, and the third for the lines representing the road centerlines and power lines.

The format of the three key files is shown below. In DOS use the Edit command and in Windows use the Notepad. Be sure to save the files with a KEY extension.

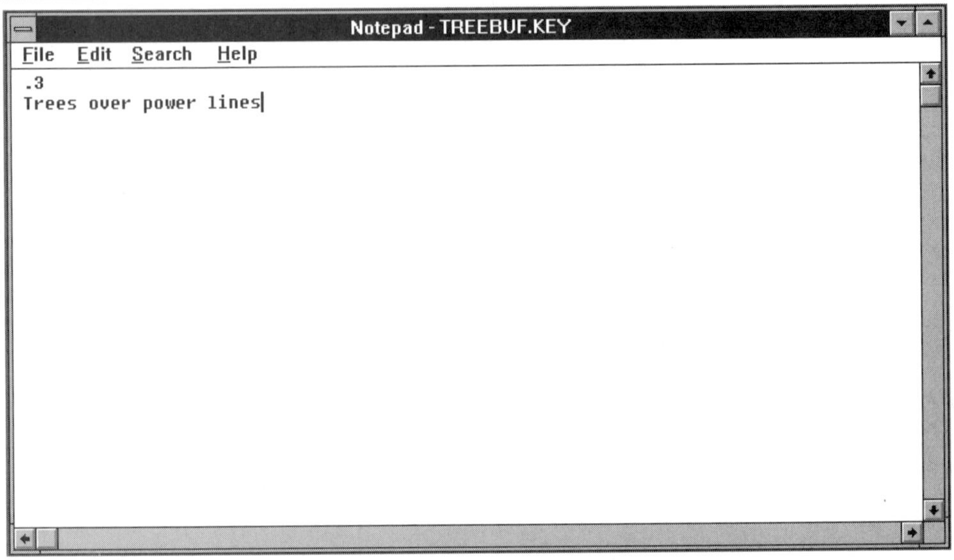

The TREES.KEY Polyshd key.

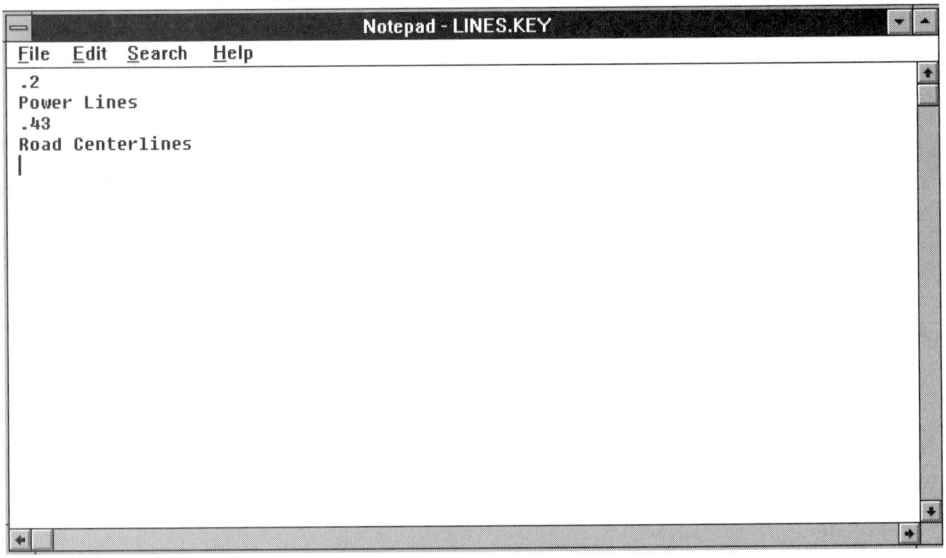

The LINES.KEY line key.

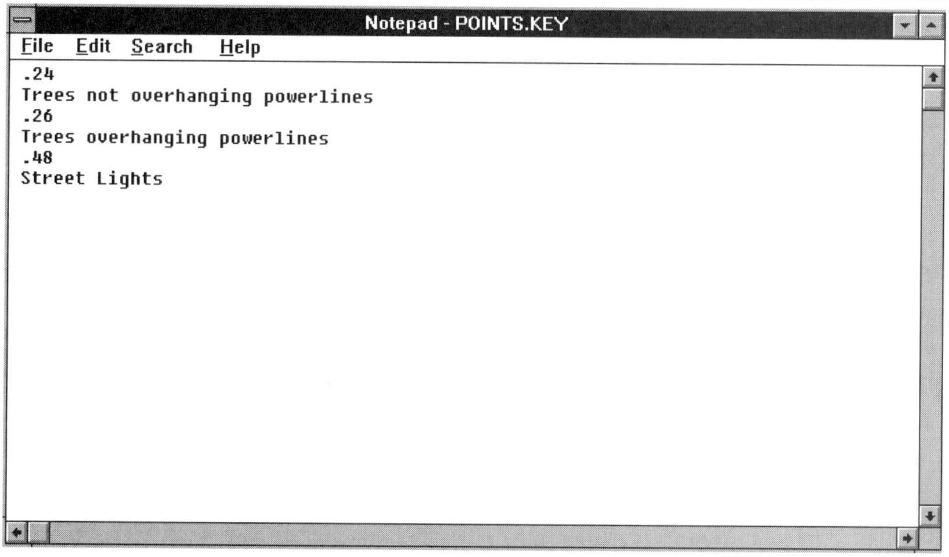
Notepad - POINTS.KEY
File Edit Search Help
.24
Trees not overhanging powerlines
.26
Trees overhanging powerlines
.48
Street Lights

The POINT.KEY Pointmrk key.

When all of the keys have been created, add them to the drawing using the Key command found on the Display menu.

```
Command: TEXTSIZE
New value for TEXTSIZE <10.0000>: 40
Command: KEY
Symbol class? (Line/Marker/Shade): S
Key file name: TREEBUF.KEY
Width of key box <150.0000>: ENTER
Height of key box <100.0000>: ENTER
Horizontal separation <25.0000>: ENTER
Vertical separation <25.0000>: ENTER
Draw key box outline? (No/<Yes>): Y
Key upper left corner (X,Y): 523530,578580
Command: KEY
Symbol class? (Line/Marker/Shade): L
Key file name: LINES.KEY
Width of key box <150.0000>: ENTER
Height of key box <100.0000>: ENTER
Horizontal separation <25.0000>: ENTER
Vertical separation <25.0000>: ENTER
Draw key box outline? (No/<Yes>): ENTER
Key upper left corner (X,Y): 523530,578320
```

```
Command: KEY
Symbol class? (Line/Marker/Shade): M
Key file name: POINTS.KEY
Width of key box <150.0000>: ENTER
Height of key box <100.0000>: ENTER
Horizontal separation <25.0000>: ENTER
Vertical separation <25.0000>: ENTER
Draw key box outline? (No/<Yes>): ENTER
Key upper left corner (X,Y): 523530,577990
```

The drawing should look like the following figure.

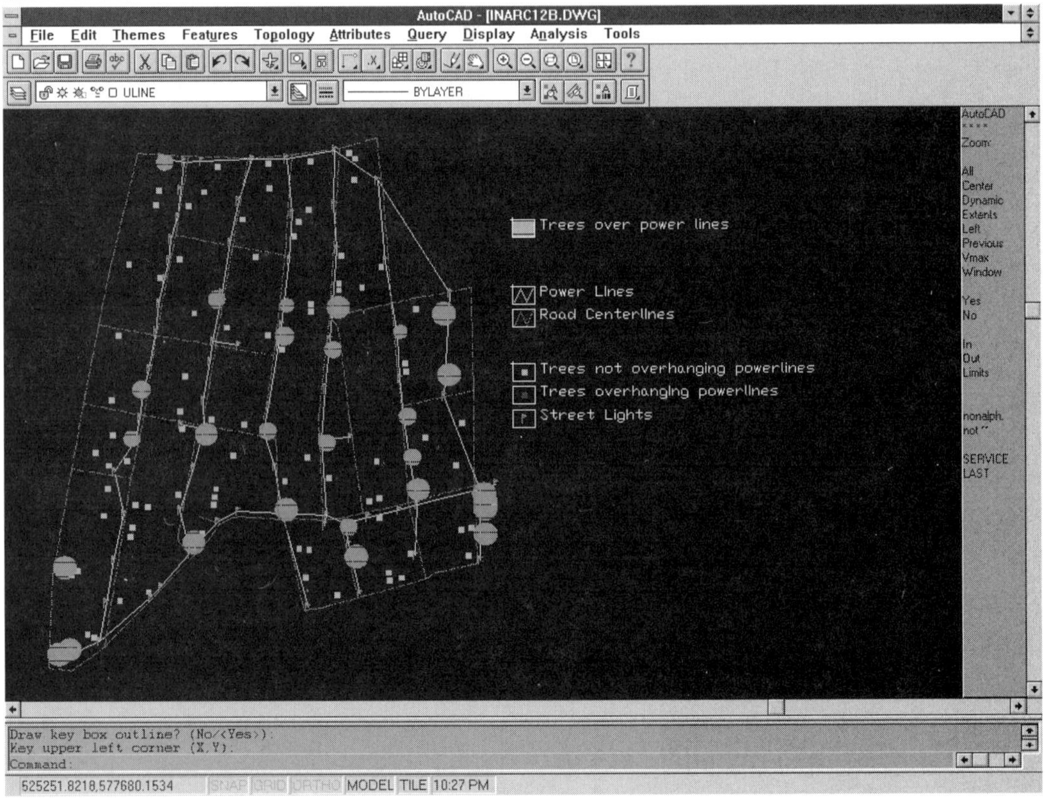

The INARC12 drawing file with legends added.

Now that the map has been completed, the TREESBUF theme, which contains only those trees that interfere with the utility lines, can be used to generate reports. An important report to generate would be one noting the location of the trees. This would be used to deploy work crews to trim or cut down trees.

To generate this report, you need to use the Near command. This command will add two items to the TREESBUF FAT. The first item is the distance between the trees and street centerlines. The second is the internal ID of the street centerlines. Once the Near command has completed, the internal ID can be used to relate the two tables. Complete this section by using earlier examples in this chapter to execute the Near command and relate the two files together.

Parcel Mapping with Lookup Tables

In Chapter 7 you created a lookup table named YEARLUT. You will use this table again to generate a thematic map. The YEARLUT lookup table is slightly different than other lookup tables in that it displays features based on a range of values, rather than a single value. When looking up a single value in a range, ArcCAD first finds a direct match. If no match is found, ArcCAD uses the next highest value. If the value being looked up is higher than the highest value in the lookup table, ArcCAD uses the last symbol in the lookup table to display the feature.

Range lookup table.

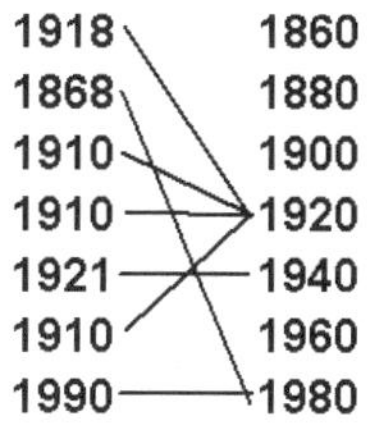

To begin this section, erase all entities from the INARC12 drawing file or use the Wblock command to save the previous map to a new drawing name.

With the drawing screen cleared, execute the Arclines command to display the street centerlines.

```
Command: ARCLINES
```

```
Theme name (?/<treesbuf>): STREETS
Display using? (Properties/Item/<Symbol>): ENTER
Line symbol number (0-100) (?/<0>): 43
```

Next, display the PARCELS into the drawing using the YEARLUT lookup table. To access the table, define a theme for it here.

```
Command: DEFTHM
```

The Defthm dialog box for the YEARLUT record theme.

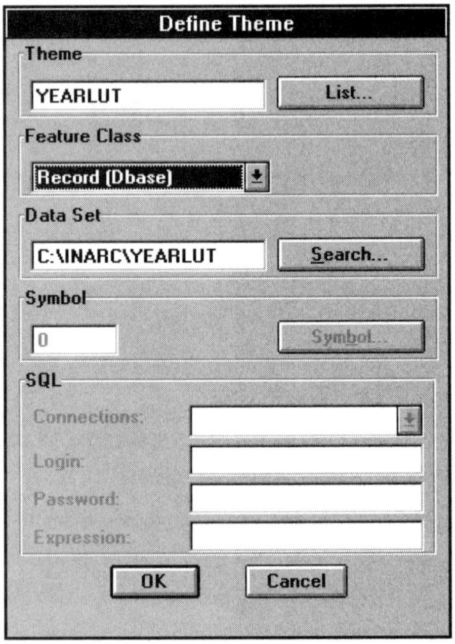

With this theme defined, execute the Polyshd command found under Polygons on the Display menu.

```
Command: POLYSHD
Theme name (?/<YEARLUT>): PARCELS
Display using? (Properties/Item/<Symbol>): I
Item (?): YRBUILT
Optional lookup table (?/<none>): YEARLUT
Coverage attribute value exceeds lookup table
Last record in lookup table will be used
```

The drawing should look like the following figure.

The INARC12 drawing file with the YEARLUT lookup table.

To complete the parcels map, create a key file using a text editor and use the Key command to create the legend. The key file and Key command are shown as follows.

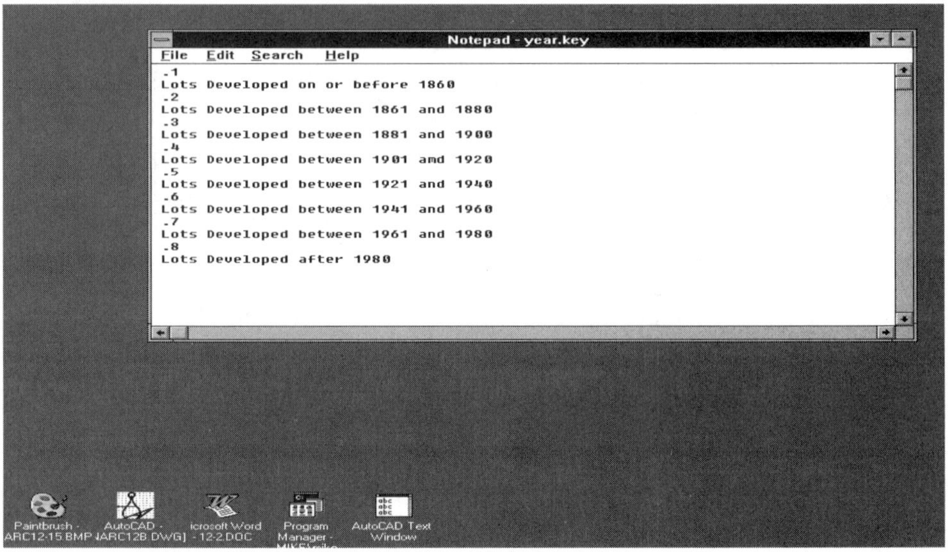

The YEAR.KEY PARCELS legend.

Notice that the last symbol in the legend is symbol number 8 for lots developed after 1980. This works as a dummy to account for any values higher than expected.

> Once the key file has been created and saved, execute the Key command found on the Display menu.

```
Command: KEY
Symbol class? (Line/Marker/Shade): S
Key file name: YEAR.KEY
Width of key box <150.0000>: ENTER
Height of key box <100.0000>: ENTER
Horizontal separation <25.0000>: ENTER
Vertical separation <25.0000>: ENTER
Draw key box outline? (No/<Yes>): ENTER
Key upper left corner (X,Y): 523700,578500
```

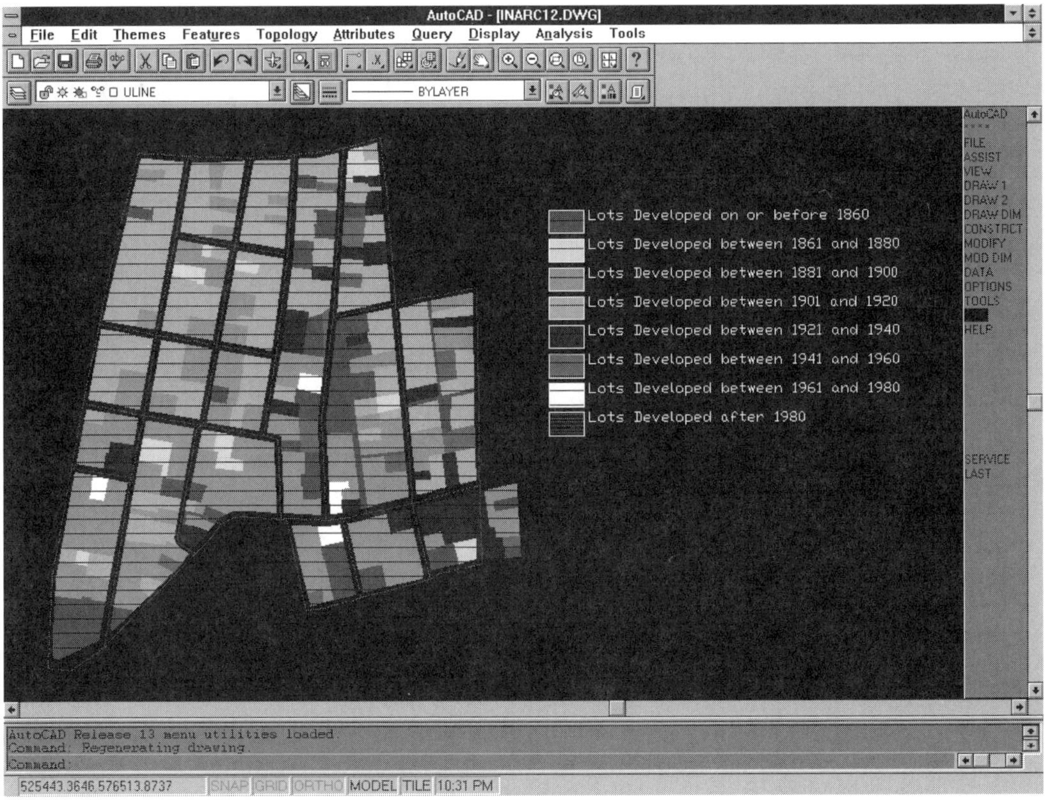

The INARC12 drawing file with key file.

Review

Chapter 12 has shown you how to create thematic maps using a series of powerful commands and features, including buffering and lookup tables. You should now be comfortable with creating and using lookup tables and creating thematic maps based on items within FATs. You should also be familiar with annotating your maps with legends. These routines allow you to automate complex mapping tasks and create professional looking drawings.

Chapter 13

Managing Your Data

The Tools Menu

ArcCAD, as a complete GIS toolkit, provides a set of commands that create and manipulate massive amounts of data. To help you deal with the management of this data, ArcCAD provides the Tools menu. Within this menu you will find commands that identify coverage types, list coverages within a specific path, and copy and rename coverages. This chapter provides an opportunity to use these commands and discusses how to manage the backup and maintenance of the data created by ArcCAD.

The commands covered in this chapter include the following:

- ❒ Listcov
- ❒ Dbdir
- ❒ Describe
- ❒ Copycov
- ❒ Renamecov
- ❒ Project

Identifying Coverage Names and Paths

The Listcov and Dbdir commands found on the Tools menu list all coverages in a specified path. The Listcov command lists point, line, and polygon coverages in a dialog box as shown in the following figure.

```
Command: LISTCOV
```

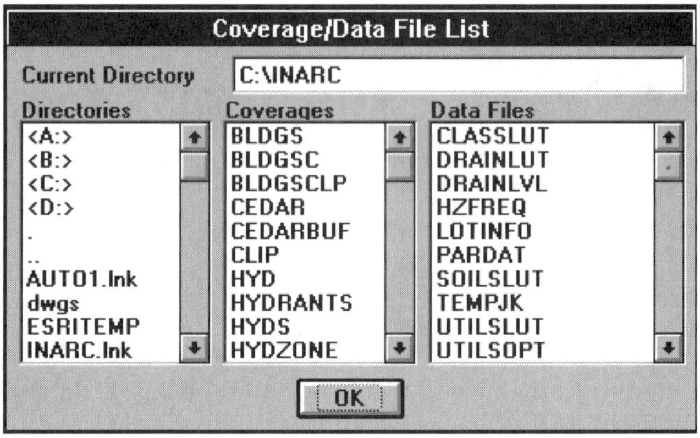

The Listcov dialog box for the INARC directory.

The Dbdir command lists all dBASE format files contained in the specified path. Both of these commands simplify the process of finding existing coverages; however, they do little to identify the contents of the coverages. For this purpose ArcCAD provides the Describe command.

Identifying Coverage Contents

The Describe command is used to display a list of information about the specified coverage. A sample listing for the STREETS coverage is shown below.

```
Command: DESCRIBE
Coverage name (?): STREETS
Coverage Description for STREETS
```

```
        ARCS                                POLYGONS
Arcs    =    70                      Polygons  =    22
Maximum Node # =  50      Polygon Topology is present.
Segments     =  94        49 bytes of Attribute Data/
                            Polygon
 128 bytes of Attribute Data/Arc

        POINTS                      SECONDARY FEATURES
Label Points  =   0        Tics      =           8
                           Annotations =         0

        TOLERANCES                          STATUS
Fuzzy   =      10.00000    There are 10 edit masks,
                             Coverage
Dangle  =      10.00000    Needs to be BUILDed or
                             CLEANed.

                    COVERAGE BOUNDARY
Xmin =      521508.8       Ymin =      576479.4
Xmax =      523408.4       Ymax =      578783.0
```

This information identifies the type of theme, point, line, polygon, as well as the size, area, and tolerances for the theme.

Copying and Renaming Coverages

The Copycov and Renamecov commands are used to copy and rename coverages, respectively. When a coverage is copied or renamed, not only are new directories created or existing ones changed, but the item names contained within the FATs are also modified.

If a coverage named STREETS is renamed, the STREETS_ and STREETS_ID items within the FAT are renamed to match the new coverage name.

NOTE: *Do not use a standard operating system command to copy a coverage. The FAT will not be updated and the coverage data will be incorrect.*

Projections

ArcCAD has the ability to project coordinates within a coverage. This ability is provided via an external DOS command that is run from the command prompt. The command accepts an input parameter file and an output parameter file. The input file defines the current projection, and the output file defines the desired projection.

Projections allow coordinates on the curved surface of the earth to be more accurately represented on a flat map. However, displaying three-dimensional information on a two-dimensional map causes some distortion. The amount of distortion can be somewhat controlled by the type of projection used. It is for this reason that it is important to determine which projection is best for your system before building the data sets.

It is not only important to determine which projection you wish to use when creating new data sets, but also when accepting data sets from outside sources. Coverages can only be projected if the user knows both the existing and the desired output projection parameters.

The concept and command input for the project command is well beyond the scope of this book. In fact, you can find more information on projects in books written specifically on the subject. You will also find information on the use of the project command in a special appendix of the *ArcCAD User's Guide.*

Review

In order to use a specific coverage, you need to know not only its location and name, but its type. The commands described here provide the means to identify the location and type of accessible coverages. You have seen not only how to identify what coverages are available for use, but also show the contents of the coverages. Coverages can also easily be copied and renamed.

Chapter 14

Advanced Features

Completing the GIS

In this chapter you will be introduced to the features of ArcCAD that allow you to take your GIS to a step above spatial and tabular data. In any GIS the spatial and tabular data make up the bulk of the information to be tracked and analyzed, but most systems also contain a substantial amount of peripheral data, such as images, scanned documents, and text files. Along with this peripheral information, most GIS users will at one point or another discover the need for functionality that doesn't exist in commercially available software packages.

This leads us to the two topics to be covered in this chapter. The first concerns the Hotlink command. This command is found on the Query menu and like most other commands is available either as a command line command or in dialog form.

Hotlinking: Integrating Other Data Sources

The Hotlink command allows you to execute functions and view image files by extracting a filename from a FAT.

Within the FAT a Hotlink item is designated. This item contains the filename to provide as the input to the function defined in the Hotlink command. For example, there is an item named IMAGE in the PARCELS

FAT. The Hotlink command can be set to IMAGE, which calls the AutoCAD rendering viewer, and the input can be set to the IMAGE item. Once set, selecting a parcel with the Hotlink command will tell ArcCAD to view the filename provided in the IMAGE item with the AutoCAD renderer. An example of this is shown in the following figure.

```
Command: DDHOTLINK
```

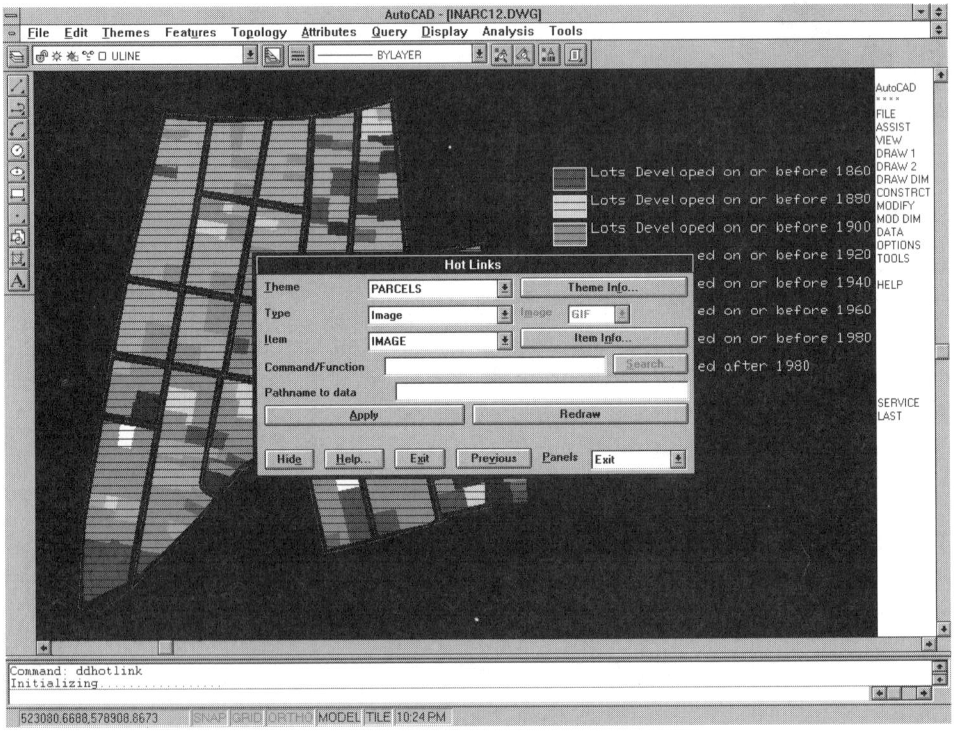

The Ddhotlink dialog box.

With the Ddhotlink dialog box set as shown, pick the Apply button and select the yellow parcel in the upper right corner of the screen. An image of the lot should be displayed in AutoCAD's rendering screen.

This is just one simple example of the power of the Hotlink command. The Hotlink command can also be used to attach text documents such as deeds or maintenance instructions, execute operating system functions, display AutoCAD slides, or execute Autolisp commands or AutoCAD scripts.

Autolisp: Creating Custom Functions

The power of the Hotlink command to develop complete data integration in ArcCAD is surpassed only by the power of the ArcCAD Autolisp and ADS extensions. ArcCAD provides two different programming interfaces to allow the customization and development of complex new functions. The two environments are the AutoCAD Autolisp programming language and a set of ADS functions for use in C programs. Both environments give the user complete access to all components of ArcCAD's spatial and tabular data.

To find out more about the Autolisp and ADS extensions, consult the *ArcCAD Programmer's Reference*, the *Autolisp Programmer's Reference*, or any of the many books on Autolisp and ADS development currently on the market.

Image Themes: Using Raster Images in ArcCAD

Image themes allow the integration of raster data with your ArcCAD-based GIS system. These themes are defined like any other theme. The only difference with image themes is that the GIS data set name for an image theme is a raster image file. ArcCAD supports a wide range of both monochrome and color raster images.

In addition to displaying raster images, ArcCAD also provides a set of utilities to register, zoom, and pan around images. All of these commands can be found on the Display menu under Images.

ArcView: Providing Mass Data Access

To not mention ArcView would leave this book incomplete. ArcView is a desktop mapping system that runs under Microsoft Windows, Mac, or UNIX. It provides a graphical point-and-shoot interface that allows the query and display of ArcCAD, PC ARC/INFO, and ARC/INFO data sets. With its graphical user interface, it is easy for the layman to query data and create map displays that can then be output to any printer supported by your operating system.

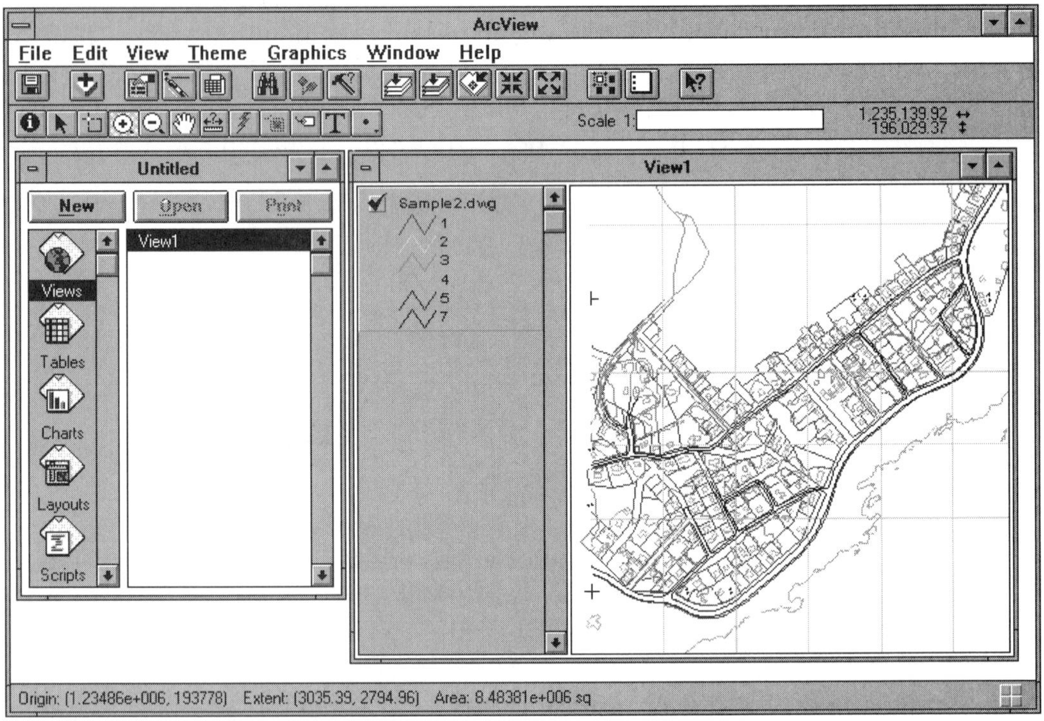

The ArcView interface in action.

With ArcView, your GIS data can be made accessible to all users who would benefit from the ability to view and perform simple analysis of the data.

Review

These more advanced features of ArcCAD allow the integration and distribution of many sources of data. Your GIS can be set up so that simple point-and-click operations provide access to varied data from photographs of features to scanned pages from manuals. With the use of the Autolisp and ADS programming extensions, a single click could allow a novice user to add new data to a coverage, such as a new subdivision to a parcel map. Although not for the new user, these programming tools can be used to simplify data gathering and analysis techniques.

You have completed your tour of ArcCAD 11.4. You should now have an understanding of how ArcCAD creates, manages, analyzes, and displays spatial and tabular datasets. You should also feel comfortable with AutoCAD's role in an ArcCAD-based GIS system. You should now have some ideas on how to expand your system beyond what is available straight out of the box.

In the Case Studies that conclude *INSIDE ArcCAD* you will see several real-world examples of ArcCAD. These examples implement all of the features presented and several of them take ArcCAD far beyond what you have seen so far.

Part IV.

Case Studies

Chapter 15

Newark Airport's Traffic Management System

Case Study

The Port Authority of New York and New Jersey developed a PC-based GIS system to provide traffic management capabilities to Newark Airport. The system combines over 20 years of traffic and accident data, coupled with AutoCAD-based drawings of all airport roads, terminal buildings, runways, sign locations, and other pertinent detail.

System Specifications

The system needed to be able to make use of existing AutoCAD drawings of the airport roadways and sign locations as well as Lotus spreadsheets of traffic and accident counts. The system needed to incorporate aerial photographs of the airport, allow immediate access to live and recorded images of the road conditions, and allow the manipulation of fiberoptic signs located throughout the airport.

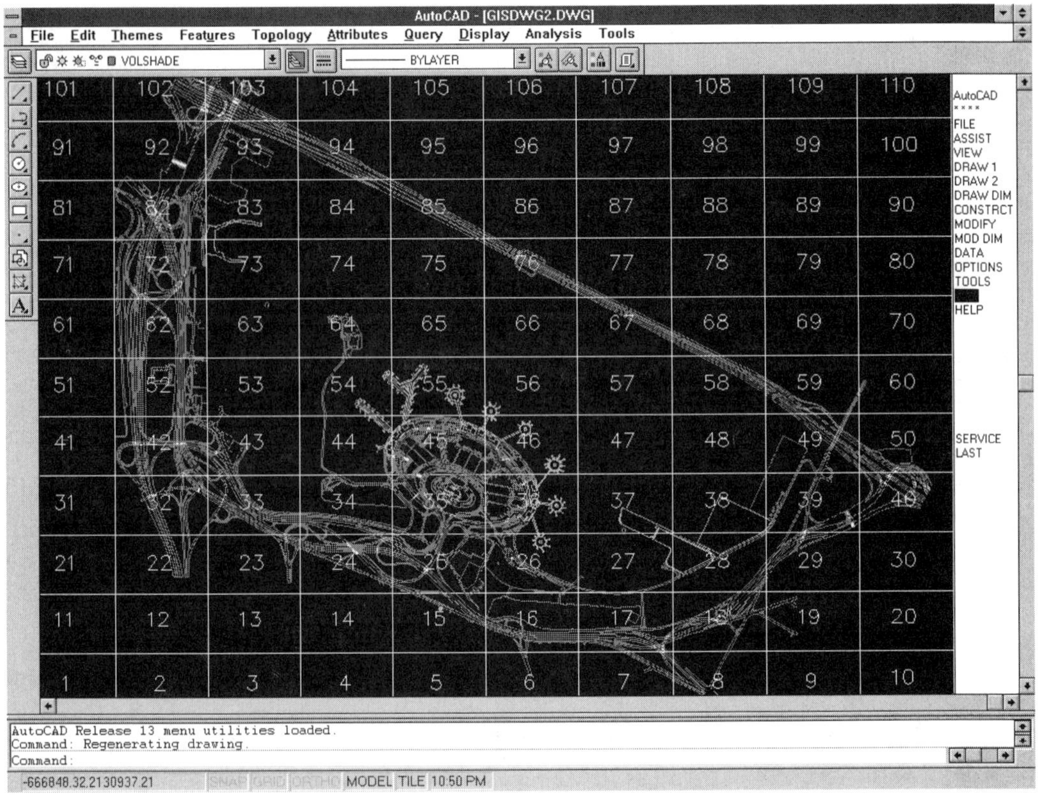

Newark Airport AutoCAD drawing.

With all of the information assembled, the system will allow the query and display of roadway polygons based on many different criteria, such as accident type, maximum volume by time of day, and maximum volume by vehicle type. The street signs will be available for query based on sign type and location. The sign symbols needed to meet state symbol standards and contain information to allow the automated retrieval of the sign's design drawing as either an image or a second AutoCAD drawing session for editing.

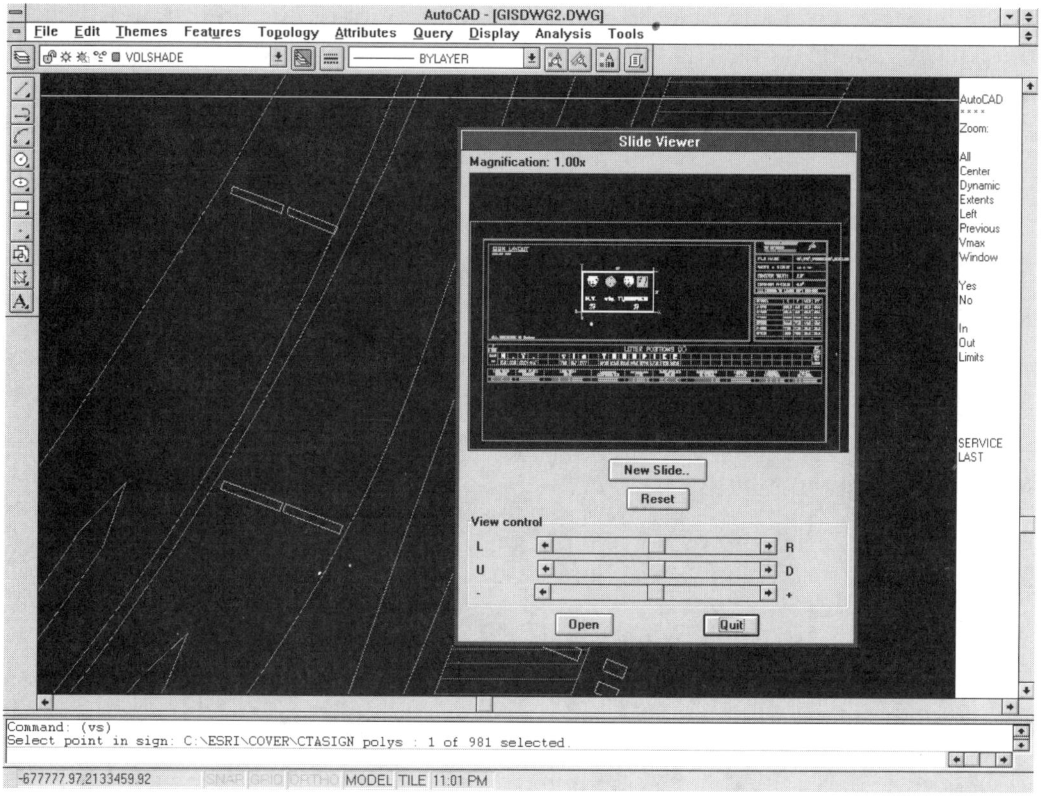

Sign design drawing.

System Design

Based on the specifications, a system based on ArcCAD was designed for Newark Airport. The system includes AutoCAD, ArcCAD, and ADE (the AutoCAD Data Extension by Autodesk, Inc.) to allow access to the original surveyed mosaic's of the airport. Animator Pro Player for Windows is used to view stored video, and Microsoft Access is used to view and manipulate the large amounts of relational data.

Designing the system involved a considerable amount of customization to the ArcCAD menu as well as the creation of several new functions created through the ArcCAD Autolisp extension. This customization included the addition of a new menu to ArcCAD tailored to the Newark Airport's specifications.

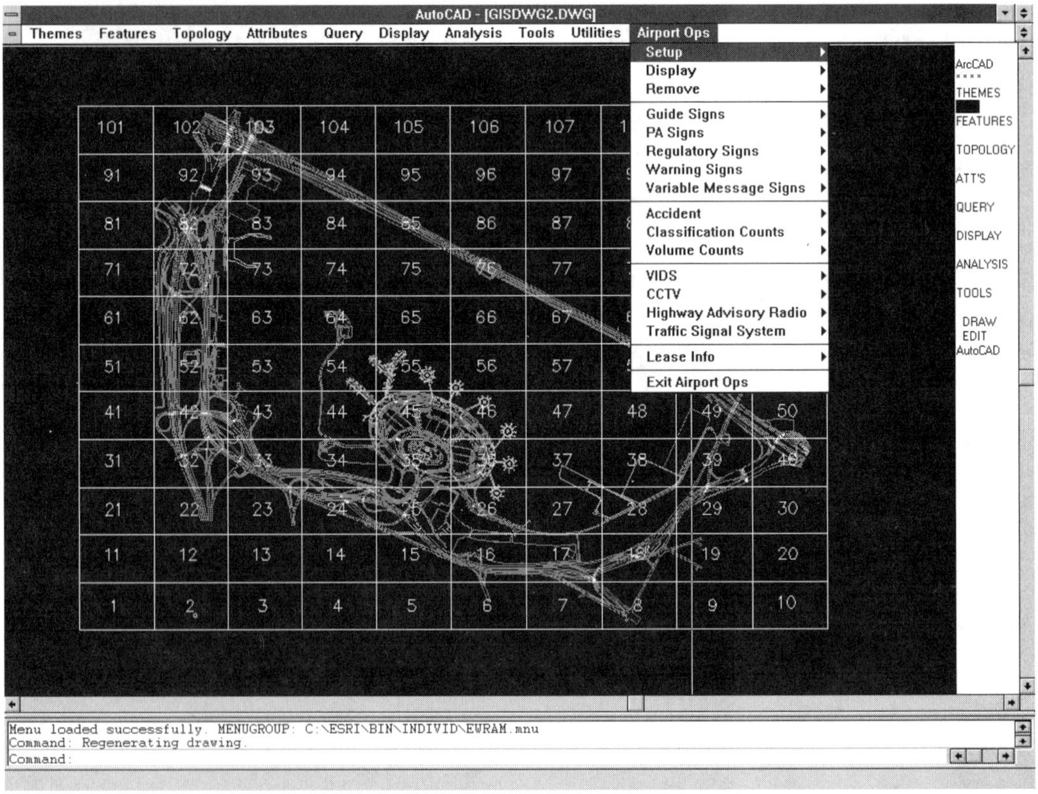

The Newark Airport Airport Ops menu.

New functions were written to allow drawing files to be imported using ADE and communication to exist seamlessly between the Access tables and the ArcCAD spatial data.

The Finished System

The Newark Airport traffic engineering staff start the system by double-clicking on a single Windows icon. The system starts Microsoft Access and ArcCAD simultaneously. ArcCAD opens with a prototype drawing that displays the Airport's sector grid, with grid numbers and all accident and volume polygons.

The aerial photograph can be turned on and off with a single menu pick.

Access to design information for the airport is provided by custom Autolisp commands and ADE. ADE allows flexible access to the detail stored in separate design drawings. All of these design drawings are stored on a central server.

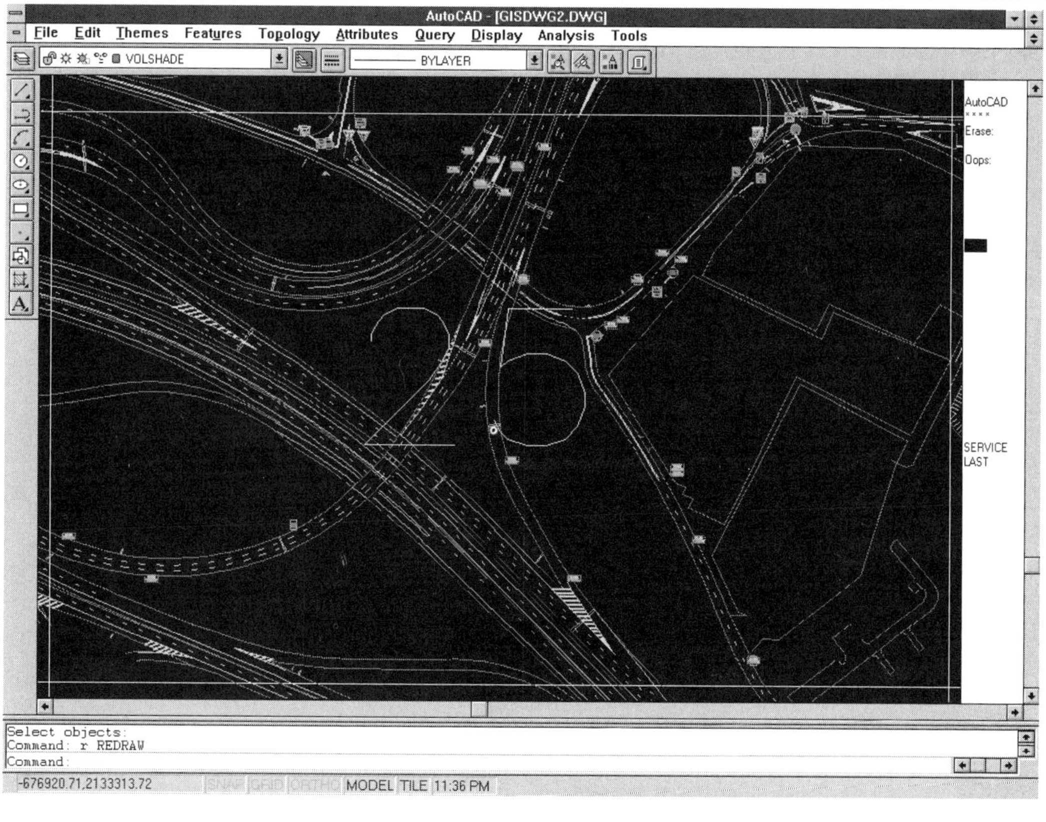

A sector with design information.

Further Autolisp customization and a series of symbol sets automate and simplify the display of the airport's traffic signs. These commands allow all of the signs or only selected signs to be displayed. The custom routines handle all layering manipulation including the removal of the signs when the display is no longer needed.

The system also allows access to many other types of data through a customized form of the Hotlink command. This data includes video segments of each of the road polygons. The video was recorded to CD, with each segment's unique name matching the corresponding road segment's number. When the user executes the View Video command and selects a polygon, the polygon number is extracted and Animator Pro Player for Windows is called with the polygon number as the filename. This same technique is used to access the sign design drawings and provide the ability to switch to Access with specific polygons selected.

With the aid of a point and shoot interface defined within Access, the user has the ability to run complex query and analysis functions.

Query dialog box in Access.

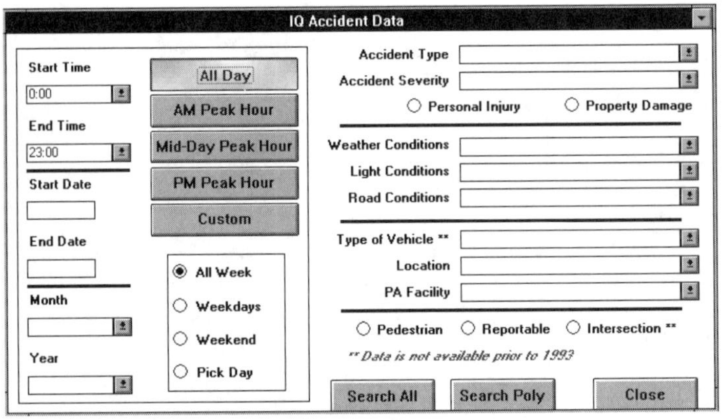

Once the queries and analysis have been completed, the results can be used to generate reports or the information can be passed back to ArcCAD for polygon shading and legend creation.

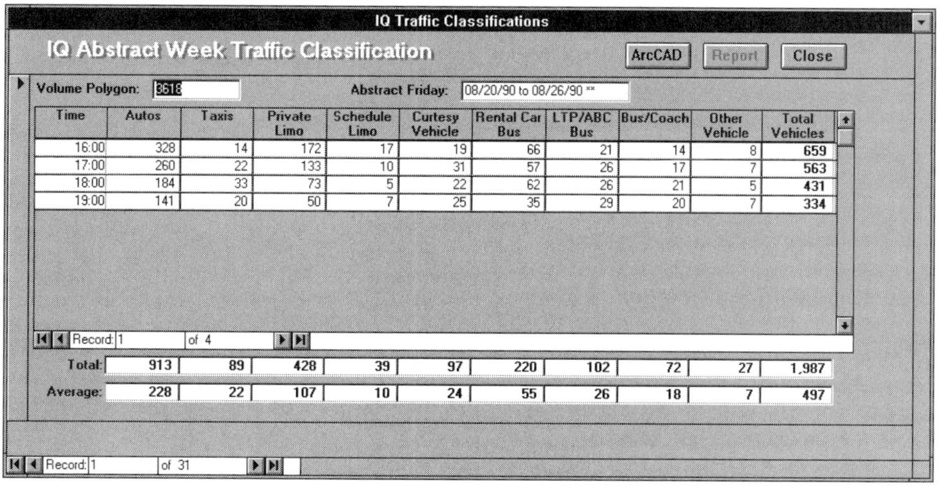

IQ Traffic Classifications

IQ Abstract Week Traffic Classification

ArcCAD | Report | Close

Volume Polygon: 3618 Abstract Friday: 08/20/90 to 08/26/90 **

Time	Autos	Taxis	Private Limo	Schedule Limo	Curtesy Vehicle	Rental Car Bus	LTP/ABC Bus	Bus/Coach	Other Vehicle	Total Vehicles
16:00	328	14	172	17	19	66	21	14	8	659
17:00	260	22	133	10	31	57	26	17	7	563
18:00	184	33	73	5	22	62	26	21	5	431
19:00	141	20	50	7	25	35	29	20	7	334
Total:	913	89	428	39	97	220	102	72	27	1,987
Average:	228	22	107	10	24	55	26	18	7	497

Record: 1 of 4

Record: 1 of 31

Report generated in Access.

Review

This case study shows the complete integration of ArcCAD under the Microsoft Windows interface, with seamless integration with other products. Not all GIS installations require this amount of customization; however, you can see the strength of the Autolisp extension and dynamic data exchange functions in this case study.

Chapter 16

The Rhode Island Department of Traffic Management Information System (TMIS)

Case Study

Introduction

Based on information outlined in the Intermodal Surface Transportation Efficiency Act (ISTEA) of 1991, the Rhode Island Department of Transportation (RIDOT) designed a GIS system to help manage design and construction projects within the state.

This system is designed to provide detailed information, allowing the RIDOT to analyze the effects of multiple highway design and reconstruction projects and make intelligent decisions on prioritizing and scheduling these projects.

The System

To design the system, RIDOT hired Beta Engineering, Inc. of Lincoln, RI, and Norwood, MA. Based on their needs analysis, Beta selected ArcCAD and AutoCAD as the software system. ArcCAD was chosen due to its ease of use, short learning curve, cost, and ability to use data from the AutoCAD and ARC/INFO system currently in use by the State of Rhode Island. The hardware used for the system was Intel 486 computers. These were chosen based on their low cost and availability.

Rhode Island TMIS logo

The Base Maps

After determining the accuracy required and the time constraints, it was decided that the system would be built around existing and readily available base maps. The data sources included the Topologically Integrated Geographic Encoding and Referencing (Tiger) files developed by the United States Geological Survey (USGS) and the Bureau of Census and the Rhode Island Geographic Information System (RIGIS) ARC/INFO files developed in a cooperative effort with RIDOT and other state and local groups.

The RIGIS files were ultimately chosen based on the appearance of the maps and the amount of data that had already been entered into the system. This base mapping included municipal boundaries, all local streets, state maintained roadways, shorelines, and bridges. RIGIS has also developed other map components including waterways, land use, soils, and wetlands that may be added to the system at a later date.

Once the determination was made to use the RIGIS ARC/INFO files, RIGIS was contacted and the necessary coverages were requested in ARC/INFO export format. Each of the export files was then imported into ArcCAD for further development.

Once imported, each of the coverages was split into tiles. Each tile represented one town. The coverages were split to reduce the size of the coverage files and increase the speed of operations.

A typical base map.

The Database

With the base maps in place, the next step in the system design was to begin compiling the project information. Information on project type, schedule, and status was compiled for all road, bridge, and signal design and construction projects. This information, tracked manually, came from numerous divisions within RIDOT. The GIS system allows the data to be compiled and entered into one integrated database that will allow for the analysis and display of a number of complex spatial relationships.

This base map attribute data was then related to an external database using a unique contract number as the link. The database was held externally to be maintained by multiple departments outside of AutoCAD. This external database contained details for each of the projects. Once the link was completed, RIDOT had a fully functional GIS to aid in its project management.

Capabilities of the System

The completed system provides RIDOT with the following features:

- A central clearinghouse for information from several sections within the department.
- A method of updating and tracking monthly project status.
- Thematic maps that can be quickly created showing the relationship between projects in any area of the state.
- Statewide maps that can be prepared for planning purposes.
- The option of ultimately linking the system to other project status reporting systems within the department.
- Comprehensive construction project information can be made available to anyone within the department using ArcCAD or ArcView.

Review

The RIDOT GIS system provides the department with a valuable tool for the overall coordination of its transportation improvement program. In addition, it provides a foundation to build upon in the development of future database management systems.

Index

A

D

E

F

M

N

O

P

T

U

More OnWord Press Titles

Pro/ENGINEER and Pro/JR. Books

INSIDE Pro/ENGINEER, 2E
Book $49.95 Includes Disk

Pro/ENGINEER Quick Reference, 2E
Book $24.95

Pro/ENGINEER Exercise Book, 2E
Book $39.95 Includes Disk

Thinking Pro/ENGINEER
Book $49.95

INSIDE Pro/JR.
Book $49.95

Interleaf Books

INSIDE Interleaf (v. 6)
Book $49.95 Includes Disk

Adventurer's Guide to Interleaf Lisp
Book $49.95 Includes Disk

The Interleaf Exercise Book
Book $39.95 Includes Disk

The Interleaf Quick Reference (v. 6)
Book $24.95

Interleaf Tips and Tricks
Book $49.95 Includes Disk

MicroStation Books

INSIDE MicroStation 5X, 3d ed.
Book $34.95 Includes Disk

MicroStation Reference Guide 5.X
Book $18.95

MicroStation Exercise Book 5.X
Book $34.95 Includes Disk
Optional Instructor's Guide $14.95

MicroStation 5.X Delta Book
Book $19.95

MicroStation for AutoCAD Users , 2E
Book $34.95

Adventures in MicroStation 3D (5.5)
Book 49.95 Includes CD

MicroStation Bible
Book $49.95
Optional Disks $49.95

Build Cell for 5.X
Software $69.95

101 MDL Commands (5.X)
Book $49.95 (4.x)
Optional Executable Disk $101.00
Optional Source Disks (6) $259.95

Managing and Networking MicroStation
Book $29.95
Optional Disk $29.95

Windows NT

Windows NT for the Technical Professional
Book $39.95

SunSoft Solaris Series

The SunSoft Solaris 2.* User's Guide
Book $29.95 Includes Disk

SunSoft Solaris 2.* for Managers and Administrators
Book $34.95

The SunSoft Solaris 2.* Quick Reference
Book $18.95

Five Steps to SunSoft Solaris 2.*
Book $24.95 Includes Disk

One Minute SunSoft Solaris Manager
Book $14.95

SunSoft Solaris 2.* for Windows Users
Book $24.95

The Hewlett Packard HP-UX Series

The HP-UX User's Guide
Book $29.95 Includes Disk

The HP-UX Quick Reference
Book $18.95

Five Steps to HP-UX
Book $24.95 Includes Disk

One Minute HP-UX Manager
Book $14.95

Other CAD

One Minute CAD Manager
Book $14.95

Manager's Guide to Computer-Aided Engineering
Book $49.95

Fallingwater in 3D Studio: A Case Study and Tutorial
Book $39.95 Includes Disk

Geographic Information Systems/ESRI

The GIS Book, 3d ed.
Book $34.95

INSIDE ARC/INFO
Book $74.95 Includes CD

ARC/INFO Quick Reference
Book $24.95

INSIDE ArcView
Book $39.95 Includes CD

ArcView Developer's Guide
Book $49.95

INSIDE ArcCAD
Book $39.95 Includes Disk

OnWord Press Distribution

End Users/User Groups/Corporate Sales

OnWord Press books are available worldwide to end users, user groups, and corporate accounts from your local bookseller or computer/software dealer, or from HMP Direct/SoftStore: call 1-800-223-6397 or 505-471-8822; fax 505-471-4424; write to High Mountain Press Direct/SoftStore, 2530 Camino Entrada, Santa Fe, NM 87505-8435, or e-mail to ORDERS@BOOKSTORE.HMP.COM.

Wholesale, Including Overseas Distribution

High Mountain Press distributes OnWord Press books internationally. For terms call 1-800-4-ONWORD or 505-471-2600; fax to 505-471-4424; e-mail to IPGORDERS@HMP.COM; or write to High Mountain Press, 2530 Camino Entrada, Santa Fe, NM 87505-8435, USA. Outside North America, call 505-471-4243.

Comments and Corrections

Your comments can help us make better products. If you find an error in our products, or have any other comments, positive or negative, we'd like to know! Please write to us at the address below or contact our e-mail address: READERS@HMP.COM.

OnWord Press

2530 Camino Entrada, Santa Fe, NM 87505-8435 USA

Chapter 9

SUPPLEMENT

This supplement to Chapter 9 should be inserted following page 204 in the accompanying book, *INSIDE ArcCAD*.

INSIDE ArcCAD®

By Michael A. DeLacey

Published by:
OnWord Press
2530 Camino Entrada
Santa Fe, NM 87505-4835

SAN 694-0269
First edition, 1995
Printed in the United States of America.

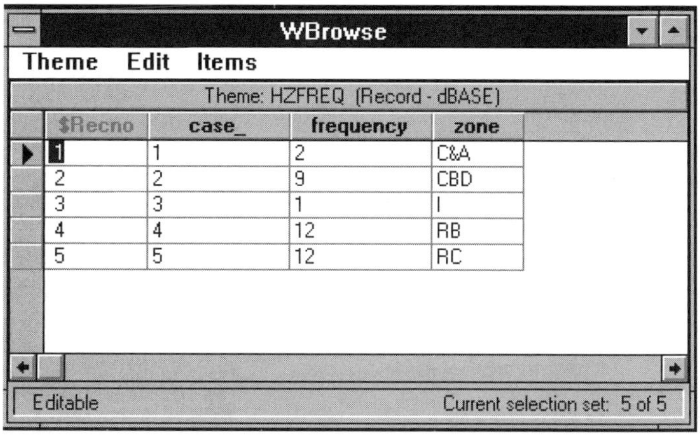

WBrowse

Theme Edit Items

Theme: HZFREQ (Record - dBASE)

$Recno	case_	frequency	zone
1	1	2	C&A
2	2	9	CBD
3	3	1	I
4	4	12	RB
5	5	12	RC

Editable Current selection set: 5 of 5

The HZFREQ record database.

Further analysis can be done for the HZFREQ theme using any application that can read dBASE III format files. For example, ArcView could be used to create a chart of the number of hydrants in each zone.

Determining Light Pole Locations

In this exercise you will use the Buffer command and a point in polygon intersection to determine which of the intersections has a light pole within 50 feet of it.

To begin this exercise, clear the parcel information from the screen by either using the Wblock command to save the information to a new drawing or using the AutoCAD Erase command to discard the information. Once the screen is clear, set the STREETS layer current and display the STREETS theme into the drawing using the Arclines command found on the Display menu.

```
Command: LA
LAYER ?/Make/Set/New/ON/OFF/Color/Ltype/Freeze/Thaw/
  LOck/Unlock: S
New current layer <PARCELS>: STREETS ?/Make/Set/New/
  ON/OFF/Color/Ltype/Freeze/Thaw/LOck/Unlock:

Command: ARCLINES
Theme name (?/<PARCELTK>): STREETS
Display using? (Properties/Item/<Symbol>): ENTER
Line symbol number (0-100) (?/<0>): 1
```

Next, create a query that will build a selection set containing all of the lights in the UTILS theme.

```
Command: RESELECT
Theme name (?/<UTILS>): RETURN
Reselect by?
  (Circle/ENtity/Expression/Polygon/Window): E
```

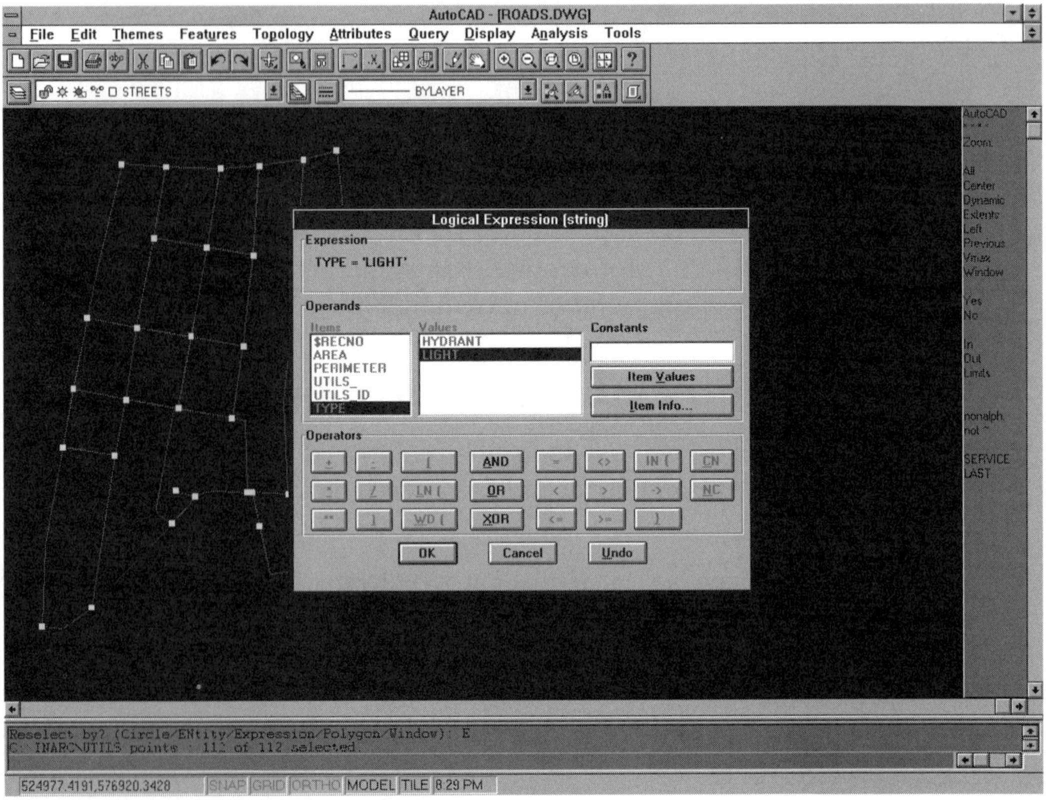

The Reselect by Expression dialog box.

Enter Type = Light in the dialog box.

```
C:\INARC\UTILS points : 112 of 112 selected.
" TYPE = 'LIGHT'" C:\INARC\UTILS points : 76 of 112
  selected.
```

Now extract the lights from the UTILS theme using the Extract command found on the Analysis menu.

```
Command: EXTRACT
Input theme (?/<UTILS>): ENTER
Output theme (?): LIGHTS
Theme LIGHTS is not defined.
```

Select O.K. from the Define Theme dialog box.

```
Defining LIGHTS...
Extracting features from UTILS to create LIGHTS.
76 of 112 records selected.
Extracting points...
Number of Points in Coverage...     112
Number of Points Reselected....      76
Creating Point attribute table.
```

The Lights theme should look as shown below.

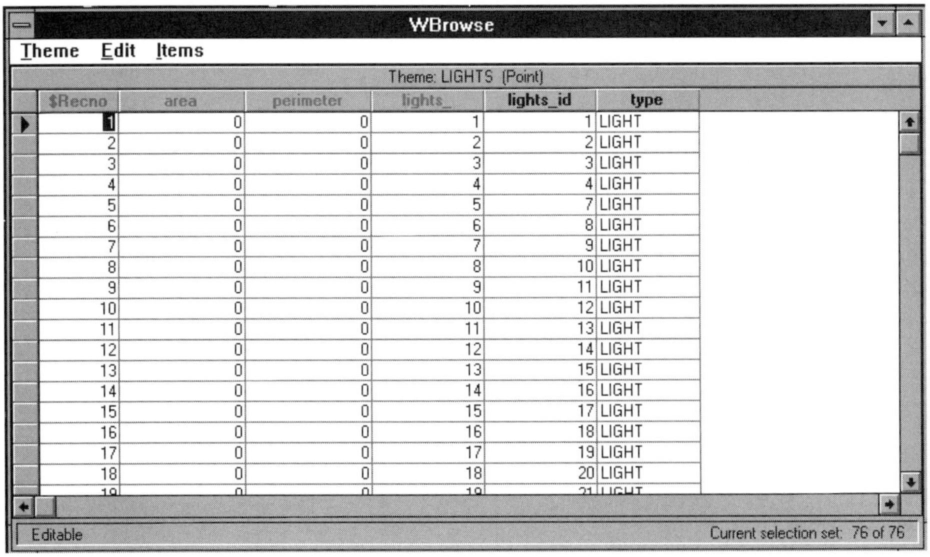

$Recno	area	perimeter	lights_	lights_id	type
1	0	0	1	1	LIGHT
2	0	0	2	2	LIGHT
3	0	0	3	3	LIGHT
4	0	0	4	4	LIGHT
5	0	0	5	7	LIGHT
6	0	0	6	8	LIGHT
7	0	0	7	9	LIGHT
8	0	0	8	10	LIGHT
9	0	0	9	11	LIGHT
10	0	0	10	12	LIGHT
11	0	0	11	13	LIGHT
12	0	0	12	14	LIGHT
13	0	0	13	15	LIGHT
14	0	0	14	16	LIGHT
15	0	0	15	17	LIGHT
16	0	0	16	18	LIGHT
17	0	0	17	19	LIGHT
18	0	0	18	20	LIGHT
19	0	0	19	21	LIGHT

The LIGHTS FAT.

To bring the light poles and intersection nodes into the drawing, use the Pointmrk command found on the Display menu under Points.

```
Command: POINTMRK
Theme name (?/<LIGHTS>): ENTER
Display using? (Properties/Item/<Symbol>): ENTER
Marker symbol number (0-100) (?/<0>): 23
```

```
Command: POINTMRK
Theme name (?/<INTER>): ENTER
Display using? (Properties/Item/<Symbol>): ENTER
Marker symbol number (0-100) (?/<0>): 25
```

The task at hand is to determine how many of the lights shown on the current drawing are within 50 feet of an intersection. To determine if any one particular intersection had a light within 50 feet of it, you could use the Reselect command using the Circle option. To analyze all of the intersections you will need to use the Buffer command. First create a 50-foot buffer around the intersections. Once the buffer theme is created, it can be overlaid with the LIGHTS theme. Any lights falling within the buffer are within 50 feet of the intersection.

Begin by buffering the INTER theme.

```
Command: BUFFER
Input theme (?/<INTER>):ENTER
Output theme (?): intbuf
Theme INTBUF is not defined.
```

Select O.K. from the Define Theme dialog box.

```
Defining INTBUF...
Buffer using? (Item/<Constant>):ENTER
Buffer distance: 50
Fuzzy tolerance <compute>:ENTER
Buffering...
Sorting...
Intersecting...
Assembling polygons...
  Sorting input file...
  Processing...
  Assigning final IDs...
  Writing arc file...
  Generating polygon report...
Labeling...
Finding inside polygons...
Dissolving...
Number of Polygons (Input,Output) =        44        39
Number of Arcs     (Input,Output) =        46        38
Creating polygon attribute table...
```

The Polys command found on the Display, Polygons menu is used to display the buffer polygons into the drawing.

```
Command: POLYS
Theme name (?/<INTBUF>):ENTER
Use closed plines? (Yes/<No>):ENTER
Line symbol number (0-100) (?/<0>): 5
```

To calculate the identity of the INTER buffers and the lights, use the Identity command found on the Analysis menu. This command will merge the two datasets by combining the two FATs. All spatial features of the input theme will be maintained and only those features of the overlay theme that coincide with the input theme will be kept.

```
Command: IDENTITY
Input theme (?/<LIGHTS>): LIGHTS
Identity theme (?): INTBUF
Output theme (?): INLIGHT
Theme INLIGHT is not defined.
```

Select O.K. from the Define Theme dialog box.

```
Defining INLIGHT...
Fuzzy tolerance <compute>: ENTER
Producing identity of LIGHTS with INTBUF to create
  INLIGHT.
Overlaying points...
Creating Point attribute table.
Item "AREA" duplicated, Join File version dropped
Item "PERIMETER" duplicated, Join File version dropped
Item "LIGHTS_" duplicated, Join File version dropped
Item "AREA" duplicated, Join File version dropped
Item "PERIMETER" duplicated, Join File version dropped
Item "INTBUF_" duplicated, Join File version dropped
```

The identity theme, INLIGHT, should look as shown below.

```
Command: WBROWSE
Theme name (?/<INLIGHT>): ENTER
```

WBrowse

Theme Edit Items

Theme: INLIGHT (Point)

$Recno	area	perimeter	inlight_	inlight_id	lights_	lights_id	type	intbuf_	intbuf_id	inside
1	0	0	1	1	49	73	LIGHT	3	2	100
2	0	0	2	2	10	12	LIGHT	1	0	1
3	0	0	3	3	1	1	LIGHT	5	4	100
4	0	0	4	4	5	7	LIGHT	7	6	100
5	0	0	5	5	8	10	LIGHT	1	0	1
6	0	0	6	6	9	11	LIGHT	6	5	100
7	0	0	7	7	48	72	LIGHT	1	0	1
8	0	0	8	8	11	13	LIGHT	1	0	1
9	0	0	9	9	6	8	LIGHT	1	0	1
10	0	0	10	10	50	74	LIGHT	1	0	1
11	0	0	11	11	51	75	LIGHT	1	0	1
12	0	0	12	12	12	14	LIGHT	1	0	1
13	0	0	13	13	13	15	LIGHT	8	7	100
14	0	0	14	14	7	9	LIGHT	9	8	100
15	0	0	15	15	52	76	LIGHT	1	0	1
16	0	0	16	16	2	2	LIGHT	10	9	100

Editable — Current selection set: 76 of 76

The INLIGHT FAT.

Notice, when displaying the INLIGHT FAT that this theme has the same number of records as the LIGHTS theme. The difference in this theme over the LIGHTS theme is that the Inside item of the INTBUF theme has been added. The Identity command was used to merge the lights and the Intbuf command to allow you to query for either the lights within 50 feet of the intersections or the lights not within 50 feet of an intersection.

To display the lights, first select all of the lights within the buffers using the Reselect command found on the Query menu.

```
Command: RESELECT
Theme name (?/<INLIGHT>): ENTER
Reselect by?
  (Circle/ENtity/Expression/Polygon/Window): E
```

The Reselect Expression dialog box.

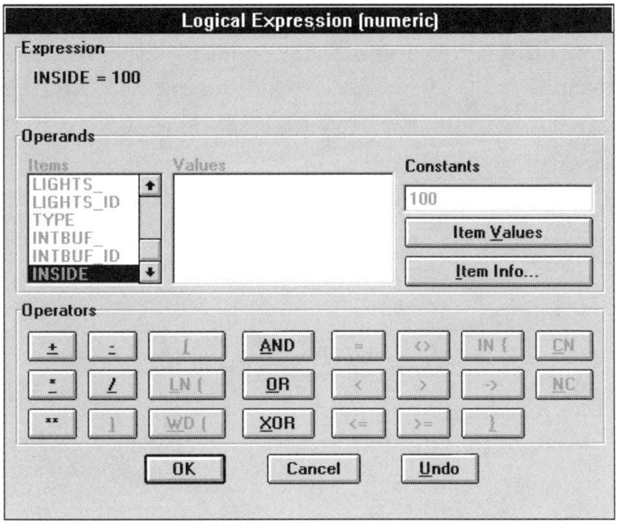

Enter INSIDE = 100 in the dialog box.

```
" INSIDE = 100" C:\INARC\INLIGHT points : 25 of 76
  selected.
```

To display the select lights into the drawing, use the Pointmrk command found on the Display, Poly's menu.

```
Command: POINTMRK
Theme name (?/<INLIGHT>): ENTER
Display using? (Properties/Item/<Symbol>): ENTER
Marker symbol number (0-100) (?/<0>): 38
```

To now switch the selection set and display the Lights outside of the buffer area, use the Nselect command found on the Query menu. The Nselect command switches the current group of records with the rest of the dataset.

```
Command: NSELECT
Theme name (?/<INLIGHT>):ENTER
C:\INARC\INLIGHT points : 51 of 76 selected.
Command: POINTMRK
Theme name (?/<INLIGHT>): ENTER
Display using? (Properties/Item/<Symbol>): ENTER
Marker symbol number (0-100) (?/<0>): 41
```

The last step in this exercise will be to remove the previous versions of the lights and the intersection nodes from the drawing. Normally, this would be very difficult considering that all of the nodes are on the same layer. To simplify this process you will use the Xselect command. The Xselect command can be found on the Query menu. This command allows the selection of AutoCAD entities based on their properties such as color or layer. To remove the pointmarks from the drawing follow the input below.

```
Command: XSELECT
Property? (Theme or
  Color/ELevation/Entity/Layer/LInetype/THickness): C
Color value: 4
Property? (Color/ELevation/Entity/Layer/LInetype/
  THickness): ENTER
41 selected.
Command: ERASE
Select objects: P
41 found
Select objects: ENTER
Command: XSELECT
Property? (Theme or
  Color/ELevation/Entity/Layer/LInetype/THickness): C
Color value: 2
Property? (Color/ELevation/Entity/Layer/LInetype/
  THickness): ENTER
76 selected.
Command: ERASE
Select objects: P
76 found
Select objects: ENTER
```

The drawing file should now look as shown in the following illustration.

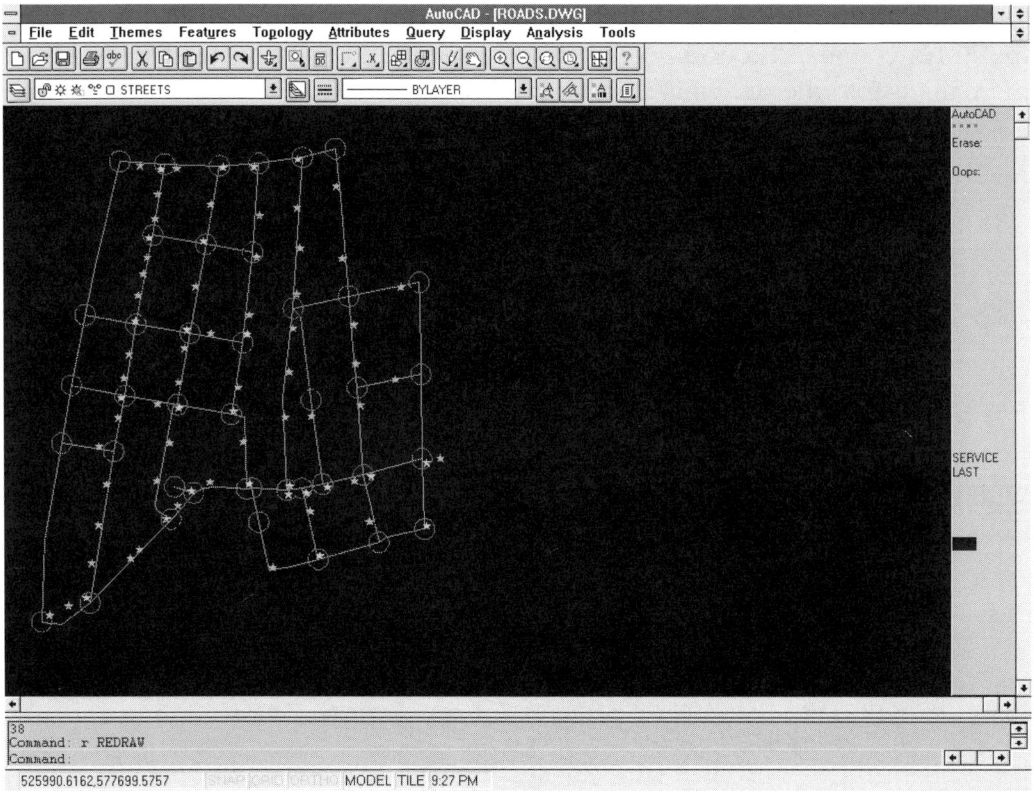

The Roads drawing with Lights displayed.

The drawing file now shows all lights within 50 feet of an intersection in green and all other lights in magenta. To store this drawing file use the AutoCAD Saveas or Wblock command now.

Managing Utility Easements

The last example in spatial analysis will be the analysis of a series of utility lines. In this project you will determine the location of utility easements across private land and the location of trees that may interfere with the utility lines.

To prepare for the exercise, clear the drawing by using the Wblock command to write the previous drawing out to a new file or use the AutoCAD Erase command to discard the previous drawing information.

Once the screen has been cleared, set the parcels layer current and use the Polys command found on the Display menu under Polys to display the parcels into the drawing.

```
Command: LA
?/Make/Set/New/ON/OFF/Color/Ltype/Freeze/Thaw/LOck/
  Unlock: S
New current layer <STREETS>: PARCELS
?/Make/Set/New/ON/OFF/Color/Ltype/Freeze/Thaw/LOck/
  Unlock:

Command: POLYS
Theme name (?/<HYFREQ>): PARCELS
Use closed plines? (Yes/<No>): ENTER
Line symbol number (0-100) (?/<0>): 4
```

With the parcels now displayed in the drawing, create a new layer named Easement and set it current.

```
Command: LA
?/Make/Set/New/ON/OFF/Color/Ltype/Freeze/Thaw/LOck/
  Unlock: M
New current layer <PARCELS>: EASEMENT
?/Make/Set/New/ON/OFF/Color/Ltype/Freeze/Thaw/LOck/
  Unlock:ENTER
```

The analysis will include the overlay of the utility lines on the parcels to reduce the set of data to only those lines that cross private land. Once the overlay has been created, a buffer representing the easement width will be created. This buffer will then be intersected with a buffer of the trees. The width item in the FAT will be used to determine the distance of the tree's buffer. The intersection of the utility buffer and the tree buffer will identify trees that interfere with the utility lines and therefore need to be maintained. Once the trees are identified, work crews may be scheduled and the maintenance tracked with the GIS system.

Begin by creating the overlay of utility lines and parcels to determine the lines that cross private land. To do so, execute the Esri_intersect command found on the Analysis menu.

```
Command: ESRI_INTERSECT
Input theme (?/<INLIGHT>): ULINE
Intersect theme (?): PARCELS
Output theme (?): PRIVLINE
```

```
Theme PRIVLINE is not defined.
Defining PRIVLINE...
Fuzzy tolerance <compute>: ENTER
Intersecting ULINE with PARCELS to create PRIVLINE.
Sorting...
Intersecting...
Assembling lines...
Creating Line attribute table.
Item "FNODE_" duplicated, Join File version dropped
Item "TNODE_" duplicated, Join File version dropped
Item "LPOLY_" duplicated, Join File version dropped
Item "RPOLY_" duplicated, Join File version dropped
Item "LENGTH" duplicated, Join File version dropped
Item "ULINE_" duplicated, Join File version dropped
Item "PARCELS_" duplicated, Join File version dropped
```

Once the intersection is complete, display the theme using the Arclines command found on the Display menu under Lines.

```
Command: ARCLINES
Theme name (?/<PRIVLINE>): ENTER
Display using? (Properties/Item/<Symbol>): ENTER
Line symbol number (0-100) (?/<0>): 5
```

This intersection has now reduced the working dataset to only utility lines that cross private land. To create the full easement across the parcels use the Buffer command found on the Analysis menu.

```
Command: BUFFER
Input theme (?/<PRIVLINE>): ENTER
Output theme (?): ULINEBUF
Theme ULINEBUF is not defined.
Defining ULINEBUF...
Buffer using? (Item/<Constant>): ENTER
Buffer distance: 10
Fuzzy tolerance <compute>: ENTER
Buffering...
Sorting...
Intersecting...
Assembling polygons...
  Sorting input file...
  Processing...
  Assigning final IDs...
  Writing arc file...
```

```
  Generating polygon report...
Labeling...
Finding inside polygons...
Dissolving...
Number of Polygons (Input,Output) =      66        16
Number of Arcs     (Input,Output) =     148        17
Creating polygon attribute table...
```

Once the buffer is created, it may be displayed into the drawing using the Polys command found on the Display menu under Polygons.

```
Command: POLYS
Theme name (?/<ULINEBUF>): ENTER
Use closed plines? (Yes/<No>): ENTER
Line symbol number (0-100) (?/<0>): 5
```

To complete the buffer process, use the Moditem command to rename the ULINEBUF's Inside item to INULINE. This is to avoid duplication of the Inside item if this buffer theme is overlaid with another buffer theme.

```
Command: MODITEM
Theme name (?/<ULINEBUF>): ENTER
Item (?): INSIDE
New item (?/<INSIDE>): INULINE
Item width <old>: ENTER
Item type (Character/Date/Numeric) <old>: ENTER
Item decimal places <old>: ENTER
```

To enable the identification of an owner with each of the easements created with the Buffer command, the utility easements must now be intersected with the PARCELS theme.

```
Command: ESRI_INTERSECT
Input theme (?/<ULINBUF>): PARCELS
Intersect theme (?): ULINEBUF
Output theme (?): UPBUF
Theme UPBUF is not defined.
Defining UPBUF...
Fuzzy tolerance <compute>: ENTER
Intersecting PARCELS with ULINEBUF to create UPBUF.
Sorting...
Intersecting...
Assembling polygons...
  Sorting input file...
  Processing...
```

```
    Assigning final IDs...
    Writing arc file...
    Generating polygon report...
  Labeling...
  Creating Polygon attribute table.
  Item "AREA" duplicated, Join File version dropped
  Item "PERIMETER" duplicated, Join File version dropped
  Item "PARCELS_" duplicated, Join File version dropped
  Item "AREA" duplicated, Join File version dropped
  Item "PERIMETER" duplicated, Join File version dropped
  Item "ULINEBUF_" duplicated, Join File version dropped
```

Intersecting the PARCELS with the ULINEBUF theme completes the preparation of the utility line easements. The UPBUF theme's FAT should look as shown below.

```
Command: WBROWSE
Theme name (?/<UPBUF>): ENTER
```

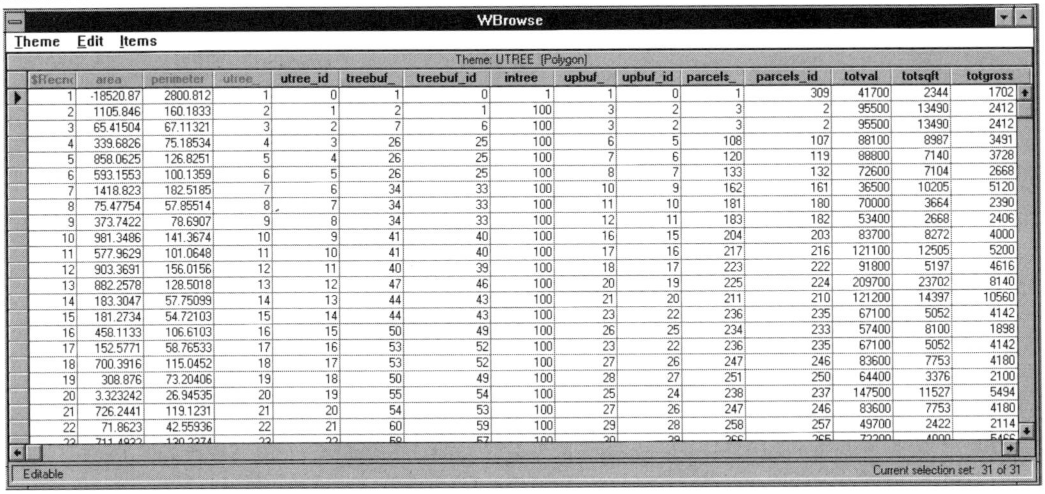

$Recno	area	perimeter	utree_	utree_id	treebuf_	treebuf_id	intree	upbuf_	upbuf_id	parcels_	parcels_id	totval	totsqft	totgross
1	-18520.87	2800.812	1	0	1	0	1	1	0	1	309	41700	2344	1702
2	1105.846	160.1833	2	1	2	1	100	3	2	3	2	95500	13490	2412
3	65.41504	67.11321	3	2	7	6	100	3	2	3	2	95500	13490	2412
4	339.6826	75.18534	4	3	26	25	100	6	5	108	107	88100	8987	3491
5	858.0625	126.8251	5	4	26	25	100	7	6	120	119	88800	7140	3728
6	593.1553	100.1359	6	5	26	25	100	8	7	133	132	72600	7104	2668
7	1418.823	182.5185	7	6	34	33	100	10	9	162	161	36500	10205	5120
8	75.47754	57.85514	8	7	34	33	100	11	10	181	180	70000	3664	2390
9	373.7422	78.6907	9	8	34	33	100	12	11	183	182	53400	2668	2406
10	981.3486	141.3674	10	9	41	40	100	16	15	204	203	83700	8272	4000
11	577.9629	101.0648	11	10	41	40	100	17	16	217	216	121100	12505	5200
12	903.3691	156.0156	12	11	40	39	100	18	17	223	222	91800	5197	4616
13	882.2578	128.5018	13	12	47	46	100	20	19	225	224	209700	23702	8140
14	183.3047	57.75099	14	13	44	43	100	21	20	211	210	121200	14397	10560
15	181.2734	54.72103	15	14	44	43	100	23	22	236	235	67100	5052	4142
16	458.1133	106.6103	16	15	50	49	100	26	25	234	233	57400	8100	1898
17	152.5771	58.76533	17	16	53	52	100	23	22	236	235	67100	5052	4142
18	700.3916	115.0452	18	17	53	52	100	27	26	247	246	83600	7753	4180
19	308.876	73.20406	19	18	50	49	100	28	27	251	250	64400	3376	2100
20	3.323242	26.94535	20	19	55	54	100	25	24	238	237	147500	11527	5494
21	726.2441	119.1231	21	20	54	53	100	27	26	247	246	83600	7753	4180
22	71.8623	42.55936	22	21	60	59	100	29	28	258	257	49700	2422	2114
23	711.4922	130.2274	23	22	58	57	100	30	29	266	265	72200	4000	5466

The UPBUF FAT.

Notice the item Parcel_ID in the UPBUF FAT. This item can be used to relate the LOTINFO theme to the UPBUF theme, thereby providing the owner's name of the parcel that the easement passes through. To exit the Wbrowse command, select Exit from the dialog box's Theme menu.

The next step in the analysis is to create a buffer for the TREES theme. The TREES theme buffer will be slightly different from the previously created buffers in that the buffer distance will be determined by a numeric item in the TREE theme's FAT. To create this buffer, follow the input below.

```
Command: BUFFER
Input theme (?/<ULINEBUF>): TREES
Output theme (?): TREEBUF
Theme TREEBUF is not defined.
Defining TREEBUF...
Buffer using? (Item/<Constant>): I
Buffer item (?): WIDTH
Optional lookup table (?/<none>): ENTER
Fuzzy tolerance <compute>: ENTER
Buffering...
Sorting...
Intersecting...
Assembling polygons...
  Sorting input file...
  Processing...
  Assigning final IDs...
  Writing arc file...
  Generating polygon report...
Labeling...
Finding inside polygons...
Dissolving...
Number of Polygons (Input,Output) =      161         81
Number of Arcs      (Input,Output) =      217         82
Creating polygon attribute table...
```

This buffer is sized based on the crown width of the trees as specified in the Width Item in the theme's FAT. By now overlaying this buffer with the utility line buffer, you can identify all trees with crown widths extending into the utility line's easement. Before doing the next overlay, use the Moditem command found on the Attributes menu under Items to change the name of the INSIDE item in the TREEBUF theme.

```
Command: MODITEM
Theme name (?/<ULINEBUF>): TREEBUF
Item (?): INSIDE
New item (?/<INSIDE>): INTREE
Item width <old>:
Item type (Character/Date/Numeric) <old>:
```

```
Item decimal places <old>:
```

With the Inside item changed, display the tree's buffer using the Polyshd command found on the Display menu under Polygons.

```
Command: POLYSHD
Theme name (?/<treebuf>): ENTER
Display using? (Properties/Item/<Symbol>): ENTER
Shade symbol number (0-100) (?/<0>): 6
```

The intersection of the TREEBUF and ULINEBUF themes are now calculated.

```
Command: ESRI_INTERSECT
Input theme (?/<TREEBUF>): ENTER
Intersect theme (?): UPBUF
Output theme (?): UTREE
Theme UTREE is not defined.
Defining UTREE...
Fuzzy tolerance <compute>: ENTER
Intersecting TREEBUF with UPBUF to create UTREE.
Sorting...
Intersecting...
Assembling polygons...
  Sorting input file...
  Processing...
  Assigning final IDs...
  Writing arc file...
  Generating polygon report...
Labeling...
Creating Polygon attribute table.
Item "AREA" duplicated, Join File version dropped
Item "PERIMETER" duplicated, Join File version dropped
Item "TREEBUF_" duplicated, Join File version dropped
Item "AREA" duplicated, Join File version dropped
Item "PERIMETER" duplicated, Join File version dropped
Item "UPBUF_" duplicated, Join File version dropped
```

To display the intersection, use the Polyshd command found on the Display menu under Polygons.

```
Command: POLYSHD
Theme name (?/<UTREE>): ENTER
Display using? (Properties/Item/<Symbol>): ENTER
Shade symbol number (0-100) (?/<0>): 1
```

The drawing should look as shown below.

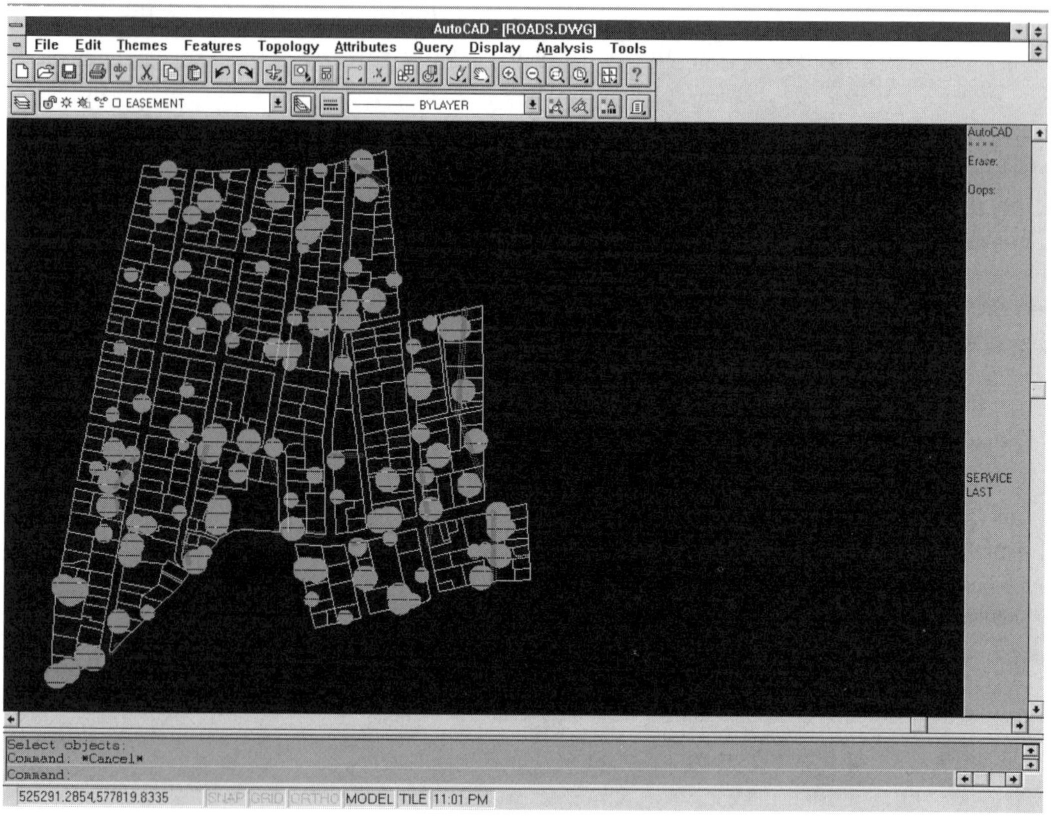

The drawing with intersection shaded.

To now generate a list of owners' names for the parcels on which tree's overhang the Utility lines, the Relate or Ddrelate command covered in earlier chapters can be used to relate the UPBUF theme to the LOTINFO theme based on the Parcels_ID item.

```
Command: RELATE
Relate? (ON/<OFF>): ON
From theme (?/<UPBUF>): ENTER
To theme (?): LOTINFO
From item (?): PARCELS_ID
To item (?): PARCELS_ID
Relate method? (Link/Ordered/Table/<LInear>): ENTER
```

The Wbrowse command can now be used to view the FAT for the UPBUF theme with its related items from the LOTINFO theme.

```
Command: WBROWSE
Theme name (?/<UPBUF>): ENTER
```

WBrowse

Theme Edit Items

Theme: UPBUF (Polygon) Related LOTINFO

$Recno	area	perimeter	upbuf_	upbuf_id	parcels_	parcels_id	totval	totsqft	totgros	image	sqftval	yrbuilt	uline	ulinebuf_i	inuline	#owner1
1	-43283.43	5336.322	1	0	1	309	41700	2344	1702		24.50	1900	1	0	1	
2	407.2666	99.71864	2	1	4	3	61400	2060	2365		25.96	1910	2	1	100	YATES BETTY
3	2862.678	337.9201	3	2	3	2	95500	13490	2412		39.59	1868	3	2	100	GARIEPY JOHN A. (TE)
4	341.0361	83.47173	4	3	8	7	65100	2591	2976		21.88	1910	4	3	100	SHUGRUE LEO V.
5	509.2158	103.6919	5	4	23	22	66900	4480	2062		32.44	1934	3	2	100	PETERS TODD A.
6	1051.441	146.2852	6	5	108	107	88100	8987	3491		25.24	1900	5	4	100	HAIDOUSIS MIHAEL (TE)
7	858.0625	126.8251	7	6	120	119	88800	7140	3728		23.82	1910	5	4	100	COULSON KATHLEEN M.
8	863.4766	126.6948	8	7	133	132	72600	7104	2668		27.21	1930	5	4	100	DOOLEY VERLEY A.
9	1972.307	237.5175	9	8	144	143	55100	16124	6238		8.83	1900	5	4	100	PAPPAS JOHN C.
10	2586.08	299.6685	10	9	162	161	36500	10205	5120		7.13	1945	5	4	100	BOWSER MARK C. (TE)
11	75.47754	57.85514	11	10	181	180	70000	3664	2390		29.29	1910	5	4	100	BIANCO FRANK J. JR. (TC
12	1085.92	150.8343	12	11	183	182	53400	2668	2406		22.19	1910	5	4	100	HOLLAND GARY W. (TE)
13	361.9551	77.08857	13	12	116	115	0	0	0		.00	0	5	4	100	LIPPISCH HANGWIND F. I
14	549.4121	119.4127	14	13	193	192	93500	5466	6205		15.07	1910	5	4	100	MACPHAIL ROBERT M. (T
15	406.752	115.0366	15	14	197	196	34600	7446	0		.00	0	5	4	100	DELUCA MICHAEL E.
16	1848.03	224.757	16	15	204	203	83700	8272	4000		20.93	1910	5	4	100	PANAGIOTES MARK TRS
17	1891.732	229.69	17	16	217	216	121100	12505	5200		23.29	1960	5	4	100	ST. JOHN JOSEPH H. JR.
18	1095.68	189.6399	18	17	223	222	91800	5197	4616		19.89	1905	6	5	100	FEDERAL NATIONAL MOF
19	309.0146	66.67334	19	18	137	136	0	0	0		.00	0	7	6	100	PROVENCAL GERARD J.

Editable

Current selection set: 43 of 4

The UPBUF FAT with LOTINFO related on Parcels_ID.

This database of information may now be used to do further analysis as well as be imported into a number of industry standard report writing products for generating form letters and reports.

Review

In this chapter you have begun to query and analyze data with ArcCAD. Commands that limit and include specific amounts of data are important for you to be able to work effectively with ArcCAD. With the Buffer, IDENTITY, ESRI_INTERSECT, and ESRI_UNION commands, you can access the information necessary for your analysis.